Changes in Water Budgets and Sediment Yields from a Hypothetical Agricultural Field as a Function of Landscape and Management Characteristics—A Unit Field Modeling Approach

By Jason L. Roth and Paul D. Capel

National Water-Quality Assessment Program

Scientific Investigations Report 2012–5203

U.S. Department of the Interior
U.S. Geological Survey

U.S. Department of the Interior
KEN SALAZAR, Secretary

U.S. Geological Survey
Marcia K. McNutt, Director

U.S. Geological Survey, Reston, Virginia: 2012

For more information on the USGS—the Federal source for science about the Earth, its natural and living resources, natural hazards, and the environment, visit http://www.usgs.gov or call 1–888–ASK–USGS.

For an overview of USGS information products, including maps, imagery, and publications, visit http://www.usgs.gov/pubprod

To order this and other USGS information products, visit http://store.usgs.gov

Suggested citation:
Roth, J.L., and Capel, P.D., 2012, Changes in water budgets and sediment yields from a hypothetical agricultural field as a function of landscape and management characteristics—A unit field modeling approach: U.S. Geological Survey Scientific Investigations Report 2012-5203, 42 p.

Foreword

The U.S. Geological Survey (USGS) is committed to providing the Nation with reliable scientific information that helps to enhance and protect the overall quality of life and that facilitates effective management of water, biological, energy, and mineral resources (http://www.usgs.gov/). Information on the Nation's water resources is critical to ensuring long-term availability of water that is safe for drinking and recreation and is suitable for industry, irrigation, and fish and wildlife. Population growth and increasing demands for water make the availability of that water, measured in terms of quantity and quality, even more essential to the long-term sustainability of our communities and ecosystems.

The USGS implemented the National Water-Quality Assessment (NAWQA) Program in 1991 to support national, regional, State, and local information needs and decisions related to water-quality management and policy (http://water.usgs.gov/nawqa). The NAWQA Program is designed to answer: What is the quality of our Nation's streams and groundwater? How are conditions changing over time? How do natural features and human activities affect the quality of streams and groundwater, and where are those effects most pronounced? By combining information on water chemistry, physical characteristics, stream habitat, and aquatic life, the NAWQA Program aims to provide science-based insights for current and emerging water issues and priorities. From 1991 to 2001, the NAWQA Program completed interdisciplinary assessments and established a baseline understanding of water-quality conditions in 51 of the Nation's river basins and aquifers, referred to as Study Units (http://water.usgs.gov/nawqa/studies/study_units.html).

In the second decade of the Program (2001–2012), a major focus is on regional assessments of water-quality conditions and trends. These regional assessments are based on major river basins and principal aquifers, which encompass larger regions of the country than the Study Units. Regional assessments extend the findings in the Study Units by filling critical gaps in characterizing the quality of surface water and groundwater, and by determining water-quality status and trends at sites that have been consistently monitored for more than a decade. In addition, the regional assessments continue to build an understanding of how natural features and human activities affect water quality. Many of the regional assessments employ modeling and other scientific tools, developed on the basis of data collected at individual sites, to help extend knowledge of water quality to unmonitored, yet comparable areas within the regions. The models thereby enhance the value of our existing data and our understanding of the hydrologic system. In addition, the models are useful in evaluating various resource-management scenarios and in predicting how our actions, such as reducing or managing nonpoint and point sources of contamination, land conversion, and altering flow and (or) pumping regimes, are likely to affect water conditions within a region.

Other activities planned during the second decade include continuing national syntheses of information on pesticides, volatile organic compounds (VOCs), nutrients, trace elements, and aquatic ecology; and continuing national topical studies on the fate of agricultural chemicals, effects of urbanization on stream ecosystems, bioaccumulation of mercury in stream ecosystems, effects of nutrient enrichment on stream ecosystems, and transport of contaminants to public-supply wells.

The USGS aims to disseminate credible, timely, and relevant science information to address practical and effective water-resource management and strategies that protect and restore water quality. We hope this NAWQA publication will provide you with insights and information to meet your needs, and will foster increased citizen awareness and involvement in the protection and restoration of our Nation's waters.

The USGS recognizes that a national assessment by a single program cannot address all water-resource issues of interest. External coordination at all levels is critical for cost-effective management, regulation, and conservation of our Nation's water resources. The NAWQA Program, therefore, depends on advice and information from other agencies—Federal, State, regional, interstate, Tribal, and local—as well as nongovernmental organizations, industry, academia, and other stakeholder groups. Your assistance and suggestions are greatly appreciated.

William H. Werkheiser
USGS Associate Director for Water

Contents

Figures

Figures—Continued

Tables

Conversion Factors, and Abbreviations and Acronyms

Conversion Factors

Inch/Pound to SI

Multiply	By	To obtain
Length		
centimeter (cm)	0.3937	inch (in.)
millimeter (mm)	0.03937	inch (in.)
meter (m)	3.281	foot (ft)
Area		
square kilometer (km^2)	247.1	acre
hectare (ha)	0.003861	square mile (mi^2)
Flow rate		
millimeter per day (mm/d)	0.03937	inch per day (in/d)
millimeter per hour (mm/h)	0.03937	inch per hour (in/h)
millimeter per year (mm/yr)	0.03937	inch per year (in/yr)
Yield		
kilogram per meter per year [kg/m)/yr]	0.6721	pound per foot per year [lb/ft)/yr]
Pressure		
pascal (Pa)	0.02088	pound per square foot (lb/ft^2)

Abbreviations and Acronyms

AMP	agricultural land management practice
NAWQA	National Water-Quality Assessment
NSERL	National Erosion Research Laboratory
USDA-ARS	U.S. Department of Agriculture – Agricultural Research Service
WEPP	Water Erosion Prediction Project

Changes in Water Budgets and Sediment Yields from a Hypothetical Agricultural Field as a Function of Landscape and Management Characteristics: A Unit Field Modeling Approach

By Jason L. Roth and Paul D. Capel

Abstract

Crop agriculture occupies 13 percent of the conterminous United States. Agricultural management practices, such as crop and tillage types, affect the hydrologic flow paths through the landscape. Some agricultural practices, such as drainage and irrigation, create entirely new hydrologic flow paths upon the landscapes where they are implemented. These hydrologic changes can affect the magnitude and partitioning of water budgets and sediment erosion. Given the wide degree of variability amongst agricultural settings, changes in the magnitudes of hydrologic flow paths and sediment erosion induced by agricultural management practices commonly are difficult to characterize, quantify, and compare using only field observations.

The Water Erosion Prediction Project (WEPP) model was used to simulate two landscape characteristics (slope and soil texture) and three agricultural management practices (land cover/crop type, tillage type, and selected agricultural land management practices) to evaluate their effects on the water budgets of and sediment yield from agricultural lands. An array of sixty-eight 60-year simulations were run, each representing a distinct natural or agricultural scenario with various slopes, soil textures, crop or land cover types, tillage types, and select agricultural management practices on an isolated 16.2-hectare field. Simulations were made to represent two common agricultural climate regimes: arid with sprinkler irrigation and humid. These climate regimes were constructed with actual climate and irrigation data. The results of these simulations demonstrate the magnitudes of potential changes in water budgets and sediment yields from lands as a result of landscape characteristics and agricultural practices adopted on them. These simulations showed that variations in landscape characteristics, such as slope and soil type, had appreciable effects on water budgets and sediment yields. As slopes increased, sediment yields increased in both the arid and humid environments. However, runoff did not increase with slope in the arid environment as was observed in the humid environment. In both environments, clayey soils exhibited the greatest amount of runoff and sediment yields while sandy soils had greater recharge and lessor runoff and sediment yield. Scenarios simulating the effects of the timing and type of tillage practice showed that no-till, conservation, and contouring tillages reduced sediment yields and, with the exception of no-till, runoff in both environments. Changes in land cover and crop type simulated the changes between the evapotransporative potential and surface roughness imparted by specific vegetations. Substantial differences in water budgets and sediment yields were observed between most agricultural crops and the natural covers selected for each environment: scrub and prairie grass for the arid environment and forest and prairie grass for the humid environment. Finally, a group of simulations was performed to model selected agricultural management practices. Among the selected practices subsurface drainage and strip cropping exhibited the largest shifts in water budgets and sediment yields. The practice of crop rotation (corn/soybean) and cover cropping (corn/rye) were predicted to increase sediment yields from a field planted as conventional corn.

Introduction

Over the past three centuries, agriculture has evolved throughout the United States, converting much of the once native landscape into what is today some of the world's most productive cropland (Sisk, 1999). A growing population with a corresponding growth in the demand for food, fiber, and fuel has resulted in the perpetual intensification of production on existing cropland as well as an expansion of agriculture onto lands that were once considered to be unarable, marginal, or once set aside for conservation purposes (Sisk, 1999; DeFries and others, 2004; Claassen and others, 2011). Agricultural cropland (excluding pasture and hay) now occupies 13 percent of the land within the United States, and production from it comprises an integral portion of the global food supply (Food and Agriculture Organization of the United Nations, 2009; Baker and Capel, 2011). The landscapes that agricultural croplands occupy have been altered from their natural pre-agricultural states to better serve the needs and demands of modern intensive agriculture. The conversion of these lands from their native states to croplands, and the subsequent land management practices and modifications implemented on them, affects the hydrologic regimes of these lands (Twine and others, 2004; Poff and others, 2006; Zhang and Schilling, 2006).

The hydrologic cycle is a function of several interdependent climatic and landscape characteristics (Chow and others, 1988). The climate of a natural landscape determines the amount and timing of precipitation and is a contributing factor in the amount of water lost through evapotranspiration from the soil and vegetative surfaces. On natural landscapes, characteristics, such as vegetation, soil type, and topography, determine the hydrologic flow paths that the water flows through following a precipitation event (Ward and Trimble, 2004). Precipitation may either be intercepted by vegetation or fall directly onto the soil. The amount of precipitation intercepted is dependent on the type of vegetation and the extent to which it covers the soil surface (Clark, 1940). Soil texture also is a major determinant of the hydrologic flow paths through the landscape; precipitation infiltrates more quickly through coarser soil textures than finer textured soils (Gupta and Larson, 1979; Brakensiek and others, 1981). Ultimately, any precipitation at intensities in excess of the combined maximum rates of interception and infiltration (for a particular combination of vegetation and soil texture) accumulates at the land surface (Kent, 1973; Chu, 1978). Landscape topography then determines whether the excess precipitation flows downgradient as runoff or, in the case of a flat land surface, continues to accumulate and pond. Generally, the steeper the slope of the land, the greater the amount of runoff; however, soil texture and vegetative cover impart a roughness to the land surface that can impede the flow of water over the surface (Stone and others, 1992). Precipitation that infiltrates into the shallow subsurface can evaporate to the atmosphere, be drawn up by the roots of vegetation and transpired back to the atmosphere, or percolate farther downward to eventually recharge groundwater.

The initial conversion to agricultural croplands altered the hydrologic regimes of natural landscapes (Mao and Cherkauer, 2009). This initial conversion often consisted of clearing the land of native vegetation and breaking up the soil to make way for future crop cultivars. These two activities altered some of the landscape's key hydrologic determinants. The removal of the native vegetation and replacement with

commonly seasonal agricultural crops altered the amount of precipitation intercepted by vegetation and the amount of water leaving the landscape through evapotranspiration (Bosch and Hewlett, 1982). Altering the amount of precipitation that fell directly on the soil and the amount, which ultimately was reintroduced to the atmosphere through evapotranspiration, altered the amount of water that flowed through the recharge and runoff flow paths. These flow paths also were affected, however, by agricultural practices, such as tillage, which alters the pore structure of the soil, an important determinant of the partitioning of precipitation between runoff and recharge in the absence of vegetation (Strudley and others, 2008). Cumulatively, these changes altered the amounts and intensity of precipitation that the vegetation and soil on the landscapes could retain prior to generating runoff, which in turn affects soil erosion.

Since the initial conversion of the natural landscape to cropland, agricultural land management practices continue to alter hydrologic determinants (water input, vegetation, soil structure and permeability, and slope). These land management practices and modifications vary both spatially and temporally. Some land management practices are spatially expansive, but seasonal in their occurrence and effects (tilling, planting, and harvesting), whereas others are more localized or more permanent (irrigation infrastructure, subsurface drains, or terraces). For example, prior to the conversion of natural landscapes to croplands, climate and its effects on vegetative growth could be considered to be the single largest variable component of the water budget. On agricultural croplands, however, an additional annual variability is introduced by recurring land management practices, such as tillage and crop type. As a result, cropland water budgets may vary from year to year. Cumulatively across the greater landscape, these induced changes of the hydrologic determinants have affected evapotranspiration from the landscape, streamflow, and groundwater recharge rates (Böhlke, 2002; Foley and others, 2004; Scanlon and others, 2005; Poff and others, 2006; Zhang and Schilling, 2006; Mao and Cherkauer, 2009).

As a consequence of the hydrologic changes associated with agricultural land management practices and landscape modifications, the geomorphologic processes of soil erosion and deposition on the landscape and sediment delivery to water bodies has been affected (Gleason and Euliss, 1983; Engstrom and others, 2009; Yan and others, 2010). Soil erosion is a natural process in landscape evolution (Lin and others, 2008). Human induced landscape changes, such as agriculture, affect the rate at which this process occurs (Montgomery, 2007), and soil erosion from croplands can have adverse effects on croplands themselves and adjacent waterways (Pimentel and others, 1987; U.S. Environmental Protection Agency, 2009). Such erosion reduces the amount of topsoil and can diminish the fertility of croplands (Pimentel and others, 1987). As the eroded soil particles are entrained in and transported in runoff and eventually reach waterways, the sediment creates turbidity in the water profile, thereby decreasing light penetration and subsequently diminishing the primary production of plant growth, which in turn can disrupt stream habitat and the food web (Newcombe and MacDonald, 1991; Quinn and others, 1992; Henley and others, 2000). Finally, sediment delivered to water bodies also can bring nutrients and other agricultural chemicals with it, which can cause still more disruption of the ecosystem (Fawcett and others, 1994; Carpenter and others, 1998; Uusitalo and others, 2001; Munn and others, 2006).

Soil erosion is affected by several hydrologic and landscape characteristics. Intense precipitation generates runoff, which dislodges and entrains sediment as it flows over the land surface (Julien, 1995). Generally, the greater the intensity and amount of runoff, the more sediment is accumulated in the runoff flow (Owoputi and Stolte, 1995). In addition, precipitation falling directly on the soil surface disturbs soil particles, making them more susceptible to being carried away by subsequent runoff. Vegetation shields the soil from direct impact by precipitation and increases the surface roughness of the land, which decreases runoff. The size and density of soil particles also is a determinant of sediment yield (Agarwal and Dickinson, 1991).

The potential effects of crop agriculture on landscape hydrology and water quality warrant consideration and quantification of the underlying hydrologic processes and their determinants. The importance of each of these inter-connected processes, however, is difficult to separate for any single location on the basis of in-situ field observations. As a result, the use of hydrologic models has proved efficient in simulating, quantifying, and comparing the effects of environmental variables on hydrologic flow paths (Woolhiser and others, 1990; Flanagan and others, 1995; Scharffenberg and Fleming, 2010). In the models, the general understanding of the movement of water is captured in theoretical or empirical relations. These relations, in their mathematical forms, are aggregated into algorithms which describe and predict the behavior of water and often other constituents, such as sediment, nutrients, or chemicals. The Water Erosion Prediction Project (WEPP) model, which was developed by the U.S. Department of Agriculture – Agricultural Research Service's (USDA-ARS) National Soil Erosion Laboratory (NSERL), is one such model that predicts hydrologic flow paths with the end goal of predicting soil erosion (Flanagan and others, 1995). The WEPP model is a process-based, distributed parameter, erosion-prediction model that has been validated with field data obtained from across the United States.

The WEPP model is capable of predicting a water budget, hydrologic flow paths (evapotranspiration, recharge, runoff, and soil water storage) and the extent of soil erosion for a field, given climate, landscape characteristics, cropping, and agricultural managements (Flanagan and others, 1995). The model requires input datasets that describe the variables of climate, slope, soil, crop type, timing and type of infield managements (planting, tillage, harvest), and any other management practices (irrigation or subsurface drainage). Using the numerical values that describe these inputs, the WEPP model simulates interrelated processes on the landscape, such as infiltration and runoff, soil compaction and erosion, and plant growth and decomposition. The WEPP model calculates daily water balances, but is capable of simulating scenarios at this temporal resolution for decades.

The WEPP model was used in this study by the U.S. Geological Survey National Water-Quality Assessment (NAWQA) Program to determine changes in the water budget and sediment yield of a hypothetical agricultural field using a unit field approach. In the study described in this report, the WEPP model was used to determine the relative effects of landscape variables and agricultural management practices on the water budget and sediment yield under two different climate regimes. An array of 60-year simulations were configured and processed in which the variables of slope, soil texture, land cover/crop type, tillage type, and selected agricultural management practices were systematically varied to isolate and, as a result, demonstrate alterations in the water budget of and erosion from the unit field. Two landscape characteristics, soil texture and slope, and three groups of agricultural management decisions of crop type, tillage type, and a broader category of agricultural management practices were considered in these simulations (fig. 1, appendix 1, tables 1A, 1B, and 1C). The ranges of the values selected for variables reflect common agricultural scenarios from across the United States. Annual statistics for the water budget components and sediment yield for each of the simulations were calculated from the model output. These statistics are briefly described and summarized in a series of tables and graphs. The results of the model simulations are used to describe the predicted absolute and relative changes to the magnitude of the water budgets, hydrologic flow paths, and sediment yield for several landscape characteristics and agricultural management practices. The results of the simulations provide improved understanding of the expected effects of specific landscape characteristics and agricultural management practices on the water budgets and sediment yields of agricultural lands.

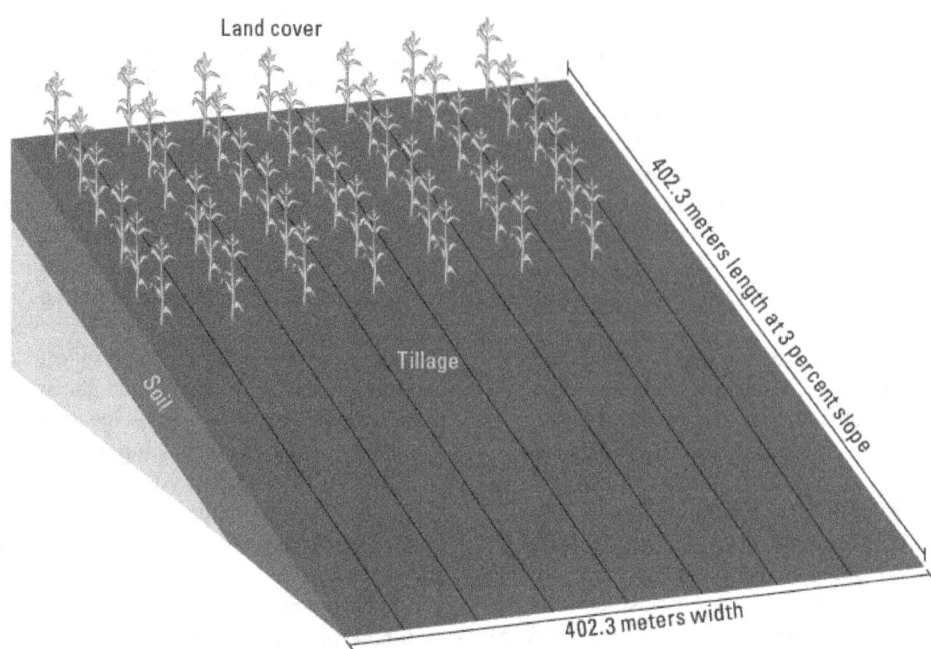

Figure 1.   The unit field and the landscape characteristics and agricultural management attributes that were isolated and varied during the course of Water Erosion Prediction Project model simulations to demonstrate their effects on the water budget and sediment yield of the unit field.

Study Methods

The WEPP model was used for simulations for this study. Details regarding the WEPP model are summarized in appendix 2.

Agricultural landscapes are mosaics of individual agricultural fields, and common landscape characteristics and decisions concerning agricultural cropland management vary at the scale of individual fields (Foster and others, 1981). The effects of these landscape characteristics and agricultural management decisions on water budgets and sediment yields were simulated using an isolated hypothetical agricultural field referred to in this report as the "unit field." In the WEPP model, the unit field was represented as a 402.3 × 402.3 m (40-acre) field of uniform hill slope profile (fig. 1). Many agricultural fields in the United States are of this dimension, which is a quarter-quarter section in the Public Land Survey System (Sisk, 1999).

Two separate climate regimes—humid and arid—were simulated using the WEPP model (fig. 2). The humid climate is representative of the Midwest Corn Belt and the arid climate is representative of much of the arid western United States. Records of daily maximum, minimum, and mean temperature and precipitation for the years 1949 through 2008 for weather stations at Greenfield, Ind., and Sunnyside, Wash. (appendix 3), were retrieved from the National Climatic Data

Center (National Climatic Data Center, 2009). These daily climate data for the weather stations were then processed using the CliGen module (Meyer, 2004) of the WEPP model to stochastically generate additional climate parameters necessary for WEPP simulations (solar radiation, wind velocity and direction, and dew point).

The sprinkler irrigation system represented in the arid climate simulations was configured on the basis of recorded irrigation data (fig. 2). The irrigation data were derived from 2 years (2003 and 2004) of daily flow rates for a drainage lateral near Sunnyside, Wash. (appendix 3). The sprinkler irrigation data required by the WEPP model includes daily depths of water delivered, hourly rates of application, and the nozzle energy of the irrigation sprinkler. The daily depth of water delivered was based on the mean daily values for the 2 years of irrigation data. To calculate the mean daily depth of water delivered by the lateral to the basin, daily lateral flows were divided by the total irrigated area of the catchment, which was 5.5 km[2]. The rate of irrigation (in millimeters per hour) was calculated on the basis of the assumption that water was delivered at a constant rate over a period of 12 hours each day. The default nozzle energy was equal to one for all irrigation dates. For a given crop or land cover, the beginning and end dates of irrigation coincided with the planting and harvesting of the crop.

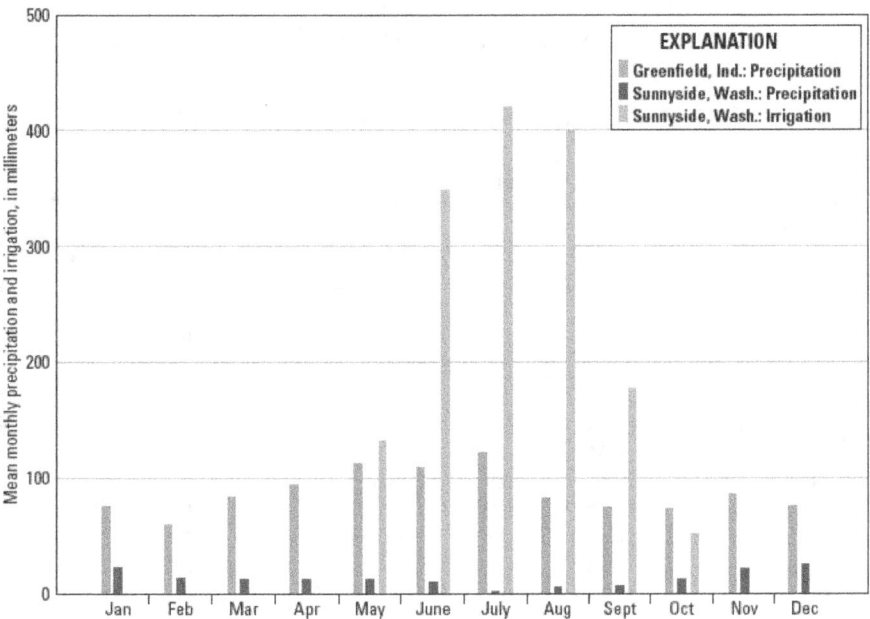

Figure 2. Mean monthly precipitation for the humid environment (Greenfield, Indiana) and precipitation and irrigation for the arid environment (Sunnyside, Washington). These climate data, along with minimum and maximum temperature data for 1949–98, are from the National Climatic Data Center (National Climatic Data Center, 2009). Irrigation data are 2-year means obtained from a drainage lateral located near Sunnyside, Washington (appendix 3).

The effects of naturally occurring variability in slope and soil type amongst agricultural fields on water budgets were considered. To demonstrate the effect of slope, a series of simulations representing constant hill slopes of 0.5, 2.0, 3.0, 6.0, 8.0, to 10.0 percent were performed. To exhibit the effects of different soil textures, a series of simulations in which soil type was varied among each of the U.S. Department of Agriculture textural classes (Natural Resources Conservation Service, 2011) was conducted to demonstrate the effect of soil type on the unit field water budget. Preconfigured soil description files for several agricultural regions within the conterminous United States, which are packaged with the WEPP distribution data, were used in the simulations (table 1). For consistency among simulations, the maximum depths of the lowest layer of the soils were set at 1.8 m; the maximum depth to which WEPP simulates recharge.

The effect of vegetative land cover on unit field water budgets was assessed using selected natural vegetation and agricultural crops in the various simulations (tables 1A, 1B, and 1C in appendix 1). Crops of corn, soybeans, wheat, and alfalfa were simulated for both climates to demonstrate the effect of crop type on the unit field water budgets. Additionally, two fallow fields, one tilled and one untilled, were simulated for each climate. For the arid fallow simulations, May 10 and October 15 were used as the beginning and end irrigation dates. Simulations also were performed using natural (pre-agricultural) land covers in order to compare pre- and post-agricultural hydrology and water budgets. In the humid climate simulations, a prairie grass and a conifer forest were simulated for this objective (table 1A), whereas a scrub cover was used for the arid climate simulation (table 1A). The arid natural vegetation simulations were run both with and without irrigation (tables 1A and 1C). Simulations run to demonstrate the effects of tillage type and timing on the unit field water budget included conditions of no till, reduced spring till, reduced fall till, conservation fall till, conventional fall till, and reduced contour fall till. The effect of a third set of agricultural management practices on the unit field water budget also was simulated. These practices consisted of annually rotating crops from corn to soybeans, planting a winter cover crop of rye after a conventional corn crop, contoured-strip cropping of corn and wheat, implementing subsurface drainage, and terracing a field such that the average slope was reduced from 8 to 3 percent.

Table 1.   U.S. Department of Agriculture soil textural classes that were included in Water Erosion Prediction Project (WEPP) model simulations and the soil configuration files selected to represent each textural class from the WEPP distribution database.

[**NRCS class**: Natural Resources Conservation Service (2011) soil textural classes. **Abbreviations**: Pa, pascal; mm/h, millimeter per hour]

Soil	NRCS class	Soil textural classes (percent)					Critical shear stress (Pa)	Vertical hydraulic conductivity (mm/h)
		Sand	Clay	Silt	Organic	Rock		
Houston	Clay	16	61	23	2	0	3.5	0.3
Eutaw	Silty clay	16	56	28	1	0	3.5	0.7
Chalmers	Silty clay loam	31	24	45	2	2	3.5	1.1
Tippo	Silt	24	13	63	0	0	3.5	2.2
Canisteo	Clay loam	41	27	32	2	4	4.2	4.2
Warden	Silt loam	36	12	53	1	1	3.5	4.4
Modesto	Loam	46	28	26	1	1	3.1	6.1
Houlka	Sandy clay loam	38	38	25	1	0	3.6	7.2
Thermo	Sandy clay	47	34	19	0	3	4.7	7.4
Thurman	Loamy sand	83	7	10	1	0	1.5	13.2
Sassafras	Sandy loam	67	13	20	1	9	2.6	13.6
Tujunga	Sand	90	3	8	0	17	2.4	17.9

In all, 68 unique, 60-year simulations were run using the WEPP model (tables 1A, 1B, and 1C). The output for each simulation contained the predicted daily depth for each of the water budget components as well as the sediment yield. These data files were then post-processed and summarized to find annual values and statistics for each of the water budget components and the sediment yield (appendix 4). For the purposes of comparison among all subgroups for each climate regime, a base scenario was included in each set of simulations. This base scenario simulation was selected to represent a common agricultural setting: 3-percent slope on a loam soil under corn production with reduced spring tillage (tables 1A, 1B, and 1C).

Annual total volumes of all water budget components and masses of sediment yield were derived from the daily output data for each simulation. From these data, mean annual values, mean annual percentages, median annual values, and annual quartiles of all water budget inflows, outflows, and sediment yields were calculated. Mean annual percentages of water budget outflows (through evapotranspiration, recharge, and runoff) were used to construct ternary plots for comparison of these components within each scenario subset. Annual median and quartile data were used to plot the relations between and variability of runoff compared to recharge and runoff compared to sediment yield.

Changes in Water Budgets and Sediment Yields as a Function of Landscape and Management Characteristics

To appreciate the effects of modern agricultural practices on field-scale water budgets, it is useful to investigate the water budgets expected on the landscape prior to agricultural development. The natural landscape evolved over many years through interactions between climate and landscape characteristics. In this subset of simulations, the pre-agricultural, or natural landscape was represented by prairie grass and forest in the humid environment and scrub in the arid environment (table 1A).

The 60-year mean annual water budget outflows for the three pre-agricultural landscape scenarios are shown at the top of figure 3: the areas of the pie graphs are proportional to the mean annual precipitation input to the water budget: 1,088 mm in the humid environment and 173 mm in the arid environment. For all of the pre-agricultural landscape scenarios evapotranspiration was the largest water budget outflow. In the arid scrub scenario, 98 percent of the precipitation that falls on the landscape is removed through evapotranspiration with runoff and recharge each accounting

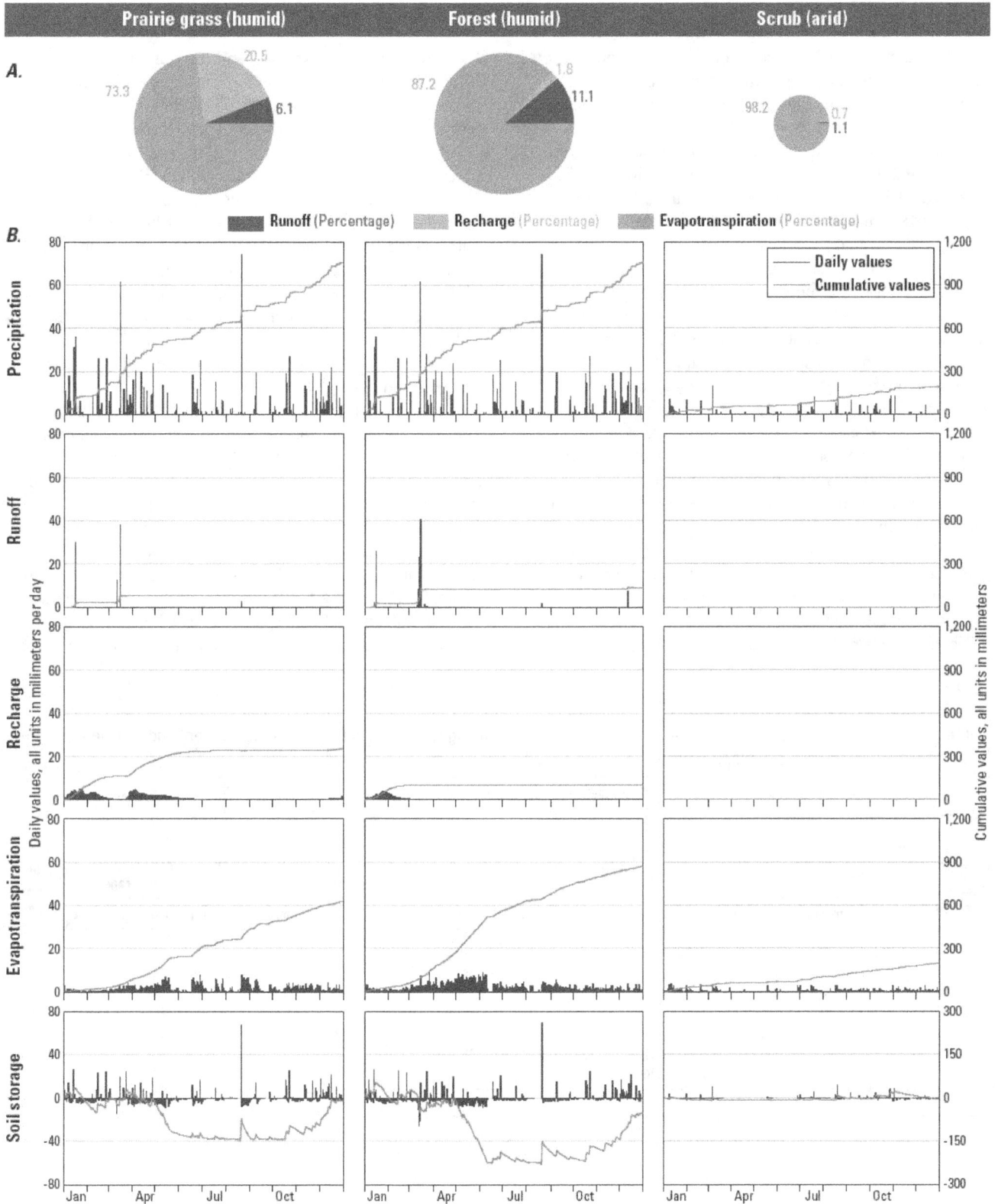

Figure 3. (A) 60-year mean annual water budget outflows with area of plot proportional to mean annual precipitation, and time-series plots (B) showing daily and cumulative values of water budget components for 2003 for pre-agricultural landscape scenarios.

for approximately 1 percent. In the humid pre-agricultural landscape scenarios, runoff and recharge comprised approximately 6 and 20 percent of the water budget outflows, respectively, on a landscape covered by prairie grass, and approximately 11 and 2 percent, respectively, in the forest scenario. The lower portion of figure 3 shows daily and cumulative values for the water budget components over the year of 2003 during which the humid environment received 1,056 mm of precipitation and the arid environment received 193 mm. The effects of individual daily precipitation events and seasonal precipitation trends have on other water budget components can be observed in figure 3. In the humid pre-agricultural landscape scenarios, runoff typically is generated only by a small number of the total annual precipitation events and most of the recharge occurs on a seasonal basis, in spring, when a the combination of a colder wetter climate and dormant vegetation decrease evapotranspiration. In the arid scenario, no runoff or recharge is predicted for the selected year. As a result of the small amount of precipitation and other climatic factors, evapotranspiration is the single water budget outflow. The following sections describe the results of the 60-year simulations in which the two landscape characteristics of slope and soil texture and the three groups of agricultural management characteristics of land cover/crop type, tillage type, and selected agricultural land management practices were systematically varied over spectrums of typical values to determine changes in the water budget and sediment yield (erosion) of the unit field.

Effects of Land Slope

Precipitation in excess of soil infiltration has an increasing tendency to run off, as the slope of the land increases (Stone and others, 1992). With this increased propensity for runoff from steeper land surfaces also comes a corresponding increased propensity for soil to be entrained and transported from the steeper lands (Julien, 1995).

The arid climate simulations with sprinkler irrigation did not indicate marked changes in the water budget components with changes in land slope (tables 2 and 3, figs. 4 and 5A). This may be a result of the delivery rate and constant intensity of the sprinkler irrigation during these simulations, which accounted for approximately 90 percent of the water budget inflows in the agricultural arid environment. However, land slope does affect the sediment yield in the arid simulations (tables 2 and 3, fig. 5B).

In the humid environment simulations, runoff increases almost linearly with slope as evapotranspiration and recharge both decrease (table 2, fig. 4B). This linear increase in runoff and decrease in recharge also is seen in figure 6A. As slope increased in the humid environment, sediment yields increased exponentially (table 3, fig. 6B). Average sediment yields were similar for equivalent slopes in both arid and humid environments, even though runoff volumes were larger in the arid environment (table 2). The variation in annual sediment yields among the different percentages of slope was much greater in the humid environment than the arid environment (table 3, figs. 5B and 6B).

Table 2.   Mean annual values and percentages of the overall water budget for each water budget component and sediment yield as a function of land slope calculated for a 60-year simulation period.

[The arid environment is representative of Sunnyside, Wash., with sprinkler irrigation, and the humid environment is representative of Greenfield, Ind. **Abbreviations**: mm/yr, millimeter per year; (kg/m)/yr, kilogram per meter per year]

Land slope (percent)	Precipitation (mm/yr)	(percent)	Irrigation (mm/yr)	(percent)	Runoff (mm/yr)	(percent)	Recharge (mm/yr)	(percent)	Evapo-transpiration (mm/yr)	(percent)	Subsurface drainage (mm/yr)	(percent)	Change in storage (mm/yr)	(percent)	Sediment yield [(kg/m)/yr]
								Arid							
0.5	173	10	1,529	90	261	15	645	38	797	47	0	0	0.1	0.0	0.01
2.0	173	10	1,529	90	261	15	645	38	797	47	0	0	0.1	0.0	0.08
3.0	173	10	1,529	90	261	15	645	38	797	47	0	0	0.1	0.0	0.60
6.0	173	10	1,529	90	260	15	644	38	797	47	0	0	0.1	0.0	6.61
8.0	173	10	1,529	90	260	15	644	38	797	47	0	0	0.1	0.0	12.54
10.0	173	10	1,529	90	260	15	644	38	797	47	0	0	0.1	0.0	19.77
								Humid							
0.5	1,088	100	0	0	81	7	234	21	773	71	0	0	0.1	0.0	0.23
2.0	1,088	100	0	0	96	9	225	21	766	70	0	0	0.1	0.0	1.20
3.0	1,088	100	0	0	102	9	222	20	763	70	0	0	0.1	0.0	2.54
6.0	1,088	100	0	0	115	11	217	20	756	70	0	0	0.1	0.0	7.80
8.0	1,088	100	0	0	122	11	213	20	753	69	0	0	0.1	0.0	11.85
10.0	1,088	100	0	0	128	12	210	19	749	69	0	0	0.1	0.0	16.30

Table 3.   Percentiles (25th, 50th, and 75th) for each water budget component and sediment yield as a function of land slope calculated for a 60-year simulation period.

[The arid environment is representative of Sunnyside, Wash., with sprinkler irrigation, and the humid environment is representative of Greenfield, Ind. **Abbreviations**: mm/yr, millimeter per year; (kg/m)/yr, kilogram per meter per year]

Land slope (percent)	Precipitation (mm/yr)			Irrigation (mm/yr)			Runoff (mm/yr)			Recharge (mm/yr)			Evapo-transpiration (mm/yr)			Subsurface drainage (mm/yr)			Change in storage (mm/yr)			Sediment yield [(kg/m)/yr]		
	25	50	75	25	50	75	25	50	75	25	50	75	25	50	75	25	50	75	25	50	75	25	50	75
Arid																								
0.5	140	169	199	1,529	1,529	1,529	213	236	259	603	642	686	792	840	855	0	0	0	−17.8	−1.3	14.2	0.00	0.01	0.01
2.0	140	169	199	1,529	1,529	1,529	213	235	259	603	642	686	792	840	855	0	0	0	−17.8	−1.3	14.2	0.00	0.01	0.07
3.0	140	169	199	1,529	1,529	1,529	213	235	259	603	642	686	792	840	855	0	0	0	−17.8	−1.3	14.2	0.12	0.25	0.67
6.0	140	169	199	1,529	1,529	1,529	212	235	259	602	642	685	792	840	855	0	0	0	−17.8	−1.3	14.2	2.67	3.67	5.70
8.0	140	169	199	1,529	1,529	1,529	212	235	259	602	642	685	792	840	855	0	0	0	−17.8	−1.3	14.2	5.44	7.49	10.27
10.0	140	169	199	1,529	1,529	1,529	212	235	259	602	642	685	792	840	855	0	0	0	−17.8	−1.3	14.2	10.00	12.62	15.98
Humid																								
0.5	952	1,064	1,217	0	0	0	31	74	116	159	240	293	715	778	840	0	0	0	−57.4	3.7	58.6	0.05	0.17	0.39
2.0	952	1,064	1,217	0	0	0	43	82	141	153	230	286	709	771	833	0	0	0	−53.4	2.5	59.4	0.20	0.69	1.98
3.0	952	1,064	1,217	0	0	0	53	89	148	152	228	284	705	768	829	0	0	0	−55.2	1 3	59.5	0.66	1.86	3.78
6.0	952	1,064	1,217	0	0	0	65	100	163	147	222	281	699	760	821	0	0	0	−57.6	−0.8	57.6	2.51	6.65	11.08
8.0	952	1,064	1,217	0	0	0	68	106	171	146	218	277	697	756	813	0	0	0	−57.5	−1.3	57.5	4.11	10.61	16.96
10.0	952	1,064	1,217	0	0	0	75	111	182	142	215	275	690	753	810	0	0	0	−57.3	−1.4	57.0	6.23	15.16	23.49

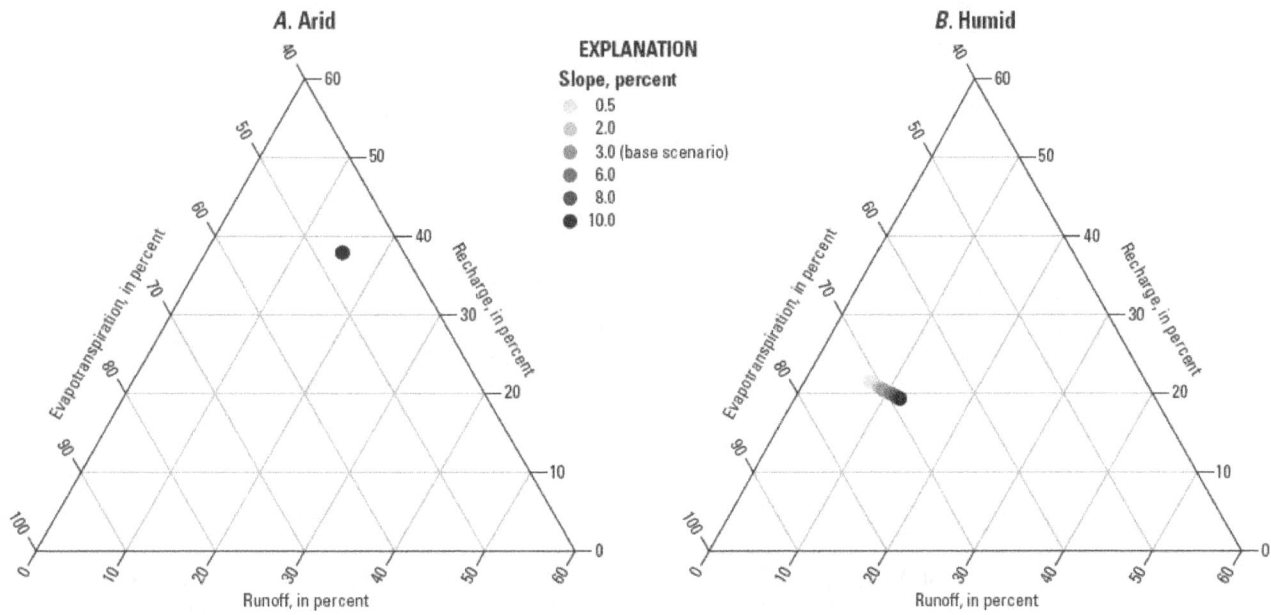

Figure 4.   60-year mean annual percentage of water budget outflows attributable to runoff, evapotranspiration, and recharge as affected by land slope for (*A*) an arid environment with sprinkler irrigation and (*B*) a humid environment. In the arid environment (*A*), variation of water budget components as a function of slope grade was minimal and thus all points appear as one point on the graph.

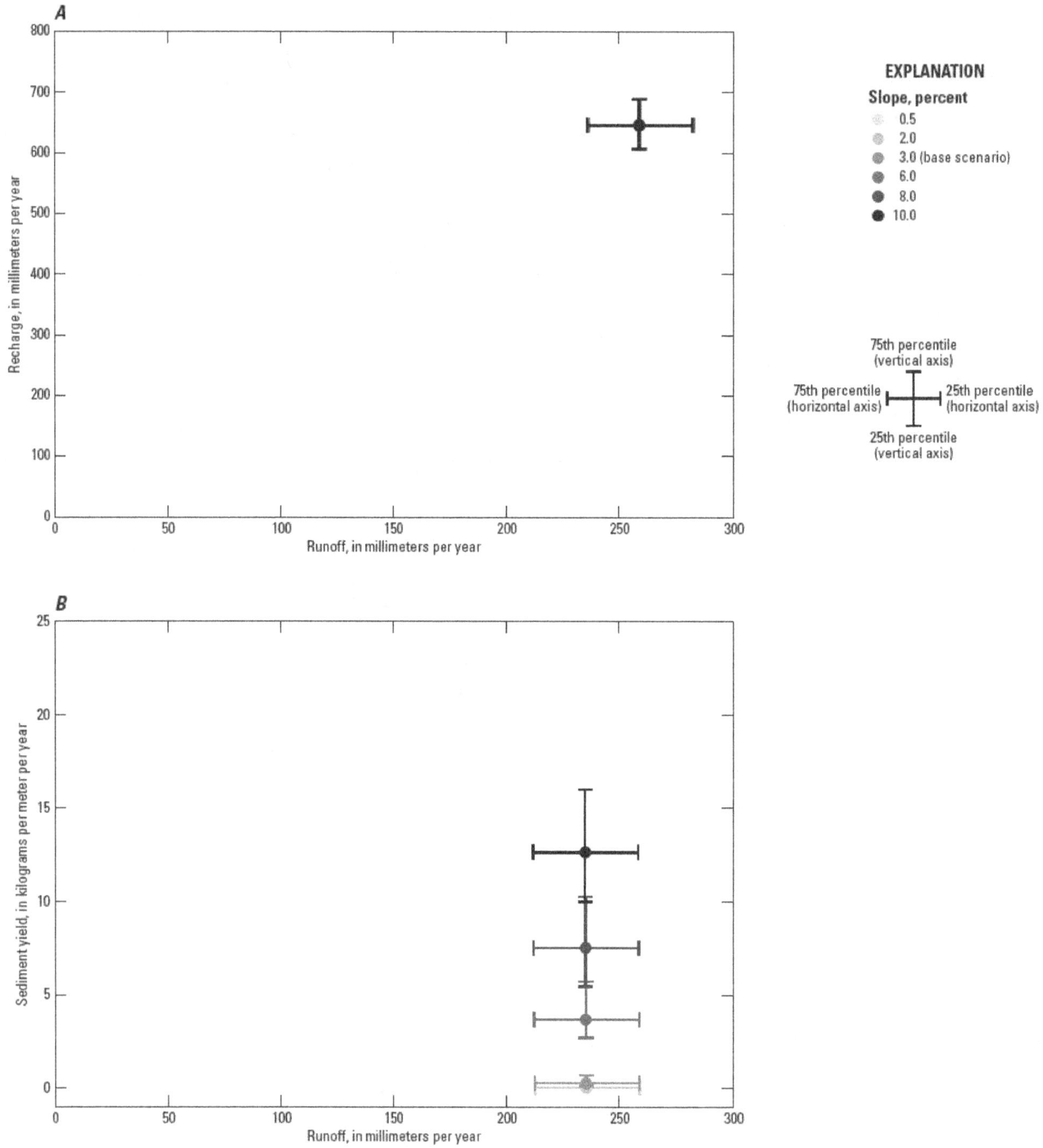

Figure 5.   60-year median values bracketed by the 25th and 75th percentiles (range bars) of relations between (*A*) runoff and recharge and (*B*) runoff and sediment yield as affected by land slope in an arid environment with sprinkler irrigation. In the arid environment (*A*), variation of water budget components as a function of slope grade was minimal and thus all points appear as on plot on the graph.

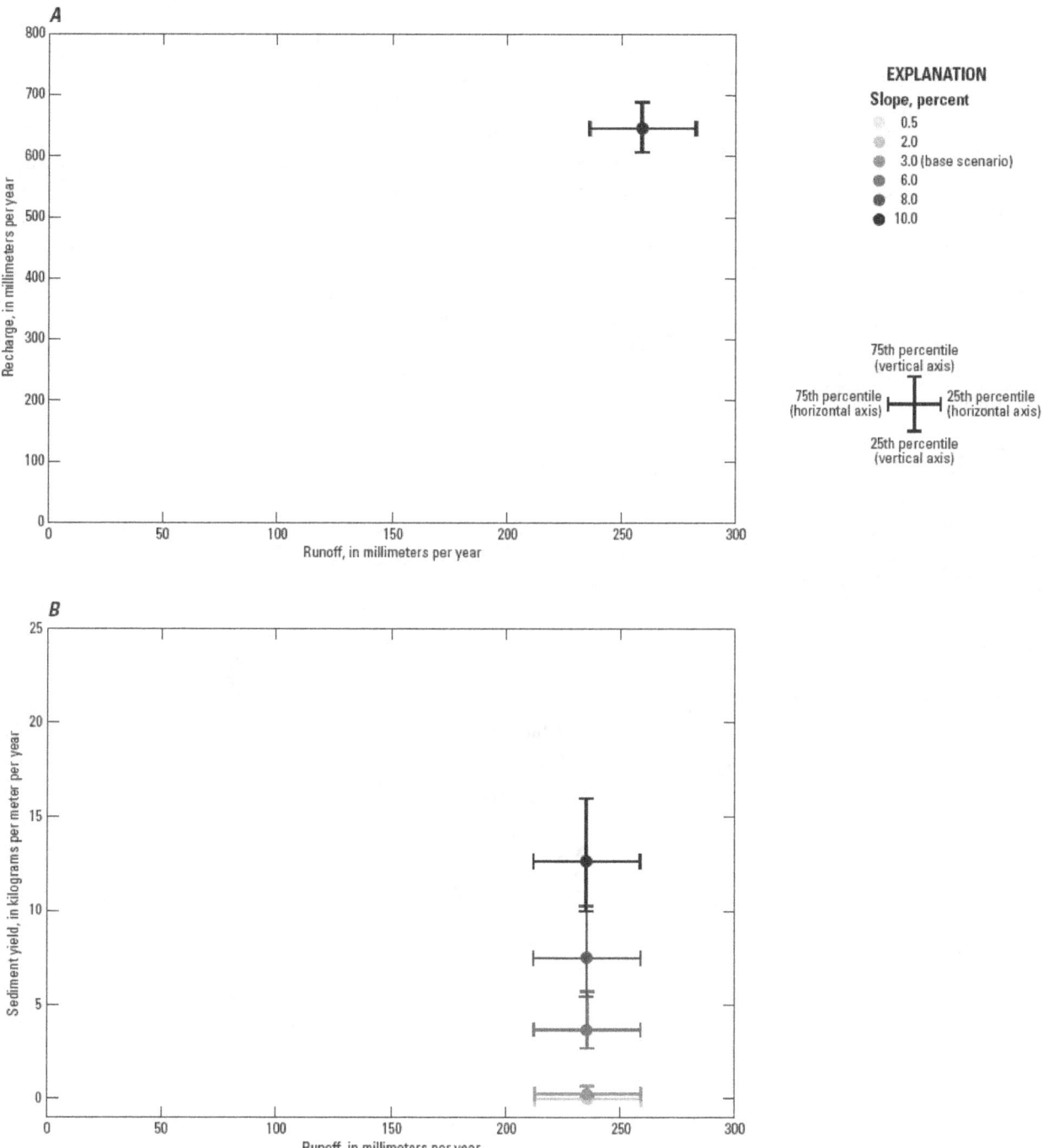

Figure 6.   60-year median values bracketed by the 25th and 75th percentiles (range bars) of relations between (*A*) runoff and recharge and (*B*) runoff and sediment yield as affected by land slope in a humid environment.

Effects of Soil Texture

The texture of a soil is determined primarily by the size distribution of its primary soil particles (Nemes and Rawls, 2004). Soil texture typically is considered with respect to three particle sizes: sand, silt, and clay. Factors, such the size, shape, and surface characteristics of the particles, affect how water is held within and flows through the soil matrix (Brakensiek and others, 1981). Soils containing a large percentage of sand particles have larger interconnected pores spaces and correspondingly less resistance to the percolation of water than do soils containing a large percentage of clay particles. Clay particles commonly form layers as a result of their surface characteristics, and the pore spaces between these layers are much smaller than the pore spaces between larger sand particles. Thus, clay-rich soils generally are less permeable than soils composed of large amounts of sand particles. However, due to the variability inherent to the pedogenesis of soils throughout the landscape, it is rare that a soil consists of solely one particle size, and commonly soils are mixtures of clay-, silt-, sand-size particles.

The effects that different soil textures have on the water budget outflows (runoff, recharge, and evapotranspiration) are shown in figure 7. Generally, for both climates, recharge increased from clayey to sandy textured soils, whereas both runoff and evapotranspiration decreased. In the arid environment, evapotranspiration was relatively constant, approximately 47 percent of total outflow for all soil textures, although there were substantial differences in the percentages of runoff and recharge (table 4 and fig. 7A). A linear decline in median annual recharge in the arid environment is predicted as soil textures transition the spectrum of sandy to clayey. In the humid environment, evapotranspiration accounted for 59–79 percent of the water outflow, and was much more variable compared to arid environment scenarios.

Sediment yields are lowest for the sandiest soils (table 5, fig. 8A). Well mixed soils in the middle of the texture spectrum have higher sediment yields than corresponding runoff values would indicate. Loam and sandy clay loam soils are as erosive as clay and silty clay soils in the arid environment (fig. 8B). In the humid environment, recharge appears to decline exponentially rather than linearly, as runoff increases and soil texture changes from sandy to clayey (table 5, fig. 9A). Sediment yields and inter-annual variability of sediment yield increase as soil textures transition from sandy to clayey in the humid environment scenarios (fig. 9B).

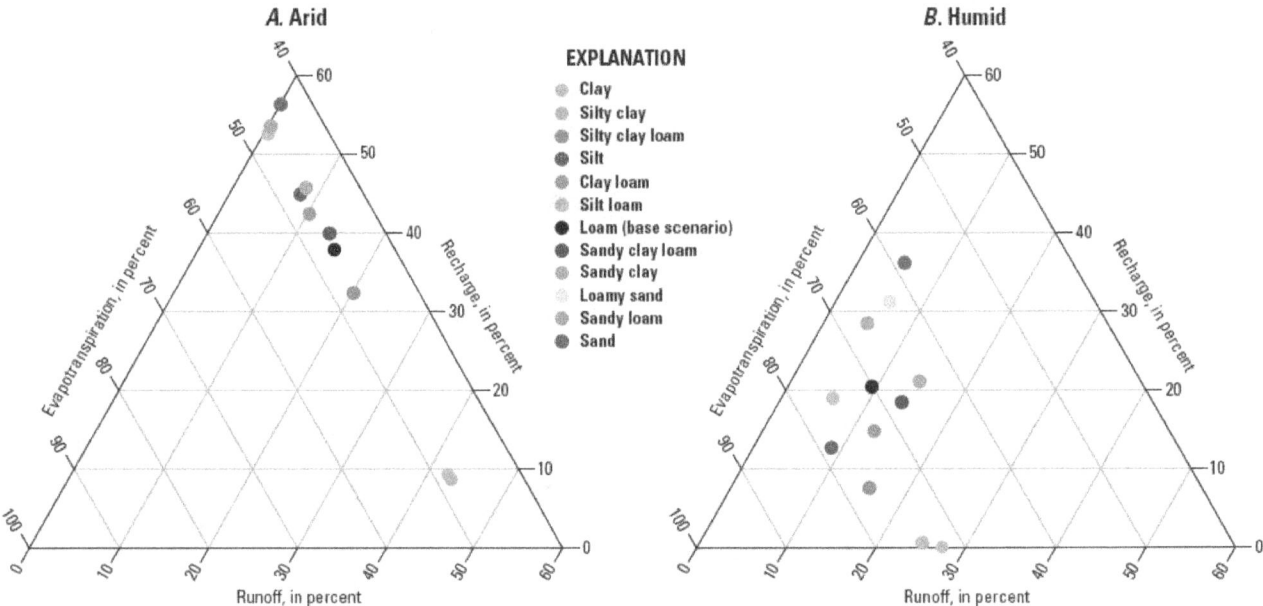

Figure 7.   60-year mean annual percentage of water budget outflows attributable to runoff, evapotranspiration, and recharge as affected by soil texture for (A) an arid environment with sprinkler irrigation and (B) a humid environment.

Table 4.   Mean annual values and percentages of overall water budget for each water budget component and sediment yield calculated for a 60-year simulation period as a function of soil texture.

[The arid environment is representative of Sunnyside, Wash., with sprinkler irrigation, and the humid environment is representative of Greenfield, Ind. **Abbreviations**: mm/yr, millimeter per year; kg/m/yr, kilogram per meter per year]

Soil texture	Precipitation		Irrigation		Runoff		Recharge		Evapo-transpiration		Subsurface drainage		Change in storage		Sediment yield
	(mm/yr)	(percent)	(mm/yr)	(percent)	(mm/yr)	(percent)	(mm/yr)	(percent)	(mm/yr)	(percent)	(mm/yr)	(percent)	(mm/yr)	(percent)	[(kg/m)/yr]
							Arid								
Clay	173	10	1,529	90	721	42	158	9	820	48	0	0	3.2	0.2	0.47
Silty clay	173	10	1,529	90	733	43	148	9	819	48	0	0	3.0	0.2	0 55
Silty clay loam	173	10	1,529	90	344	20	551	32	808	47	0	0	0.0	0.0	0.40
Silt	173	10	1,529	90	136	8	764	45	803	47	0	0	−0.5	0.0	0 20
Clay loam	173	10	1,529	90	174	10	722	42	806	47	0	0	0.2	0.0	0 11
Silt loam	173	10	1,529	90	9	1	896	53	799	47	0	0	−1.2	−0.1	0.01
Loam	173	10	1,529	90	261	15	645	38	797	47	0	0	0.1	0.0	0.60
Sandy clay loam	173	10	1,529	90	233	14	679	40	788	46	0	0	1.8	0.1	0 54
Sandy clay	173	10	1,529	90	139	8	778	46	784	46	0	0	0.6	0.0	0 11
Loamy sand	173	10	1,529	90	7	0	913	54	786	46	0	0	−3.9	−0.2	0.01
Sandy loam	173	10	1,529	90	6	0	912	53	787	46	0	0	−1.9	−0.1	0.00
Sand	173	10	1,529	90	1	0	961	56	744	44	0	0	−4.0	−0.2	0.00
							Humid								
Clay	1,088	100	0	0	297	27	1	0	788	73	0	0	1.0	0.1	4 21
Silty clay	1,088	100	0	0	270	25	7	1	809	74	0	0	1.3	0.1	3 97
Silty clay loam	1,088	100	0	0	170	16	82	8	837	77	0	0	−1.7	−0.2	2.81
Silt	1,088	100	0	0	96	9	138	13	856	79	0	0	−2.4	−0.2	2.45
Clay loam	1,088	100	0	0	136	12	161	15	792	73	0	0	−0.7	−0.1	2 14
Silt loam	1,088	100	0	0	64	6	206	19	820	75	0	0	−2.5	−0.2	2.40
Loam	1,088	100	0	0	102	9	222	20	763	70	0	0	0.1	0.0	2 54
Sandy clay loam	1,088	100	0	0	149	14	201	18	737	68	0	0	1.4	0.1	4 31
Sandy clay	1,088	100	0	0	156	14	229	21	701	65	0	0	0.6	0.1	2 97
Loamy sand	1,088	100	0	0	66	6	341	31	684	63	0	0	−3.0	−0.3	1.08
Sandy loam	1,088	100	0	0	53	5	310	28	726	67	0	0	−1.6	−0.1	0.76
Sand	1,088	100	0	0	57	5	395	36	639	59	0	0	−2.7	−0.2	0.48

Table 5.   Percentiles (25th, 50th, and 75th) for each water budget component and sediment yield calculated for a 60-year simulation period as a function of soil texture.

[The arid climate is representative of Sunnyside, Wash., with sprinkler irrigation, and the humid climate is representative of Greenfield, Ind. **Abbreviations**: mm/yr, millimeter per year; (kg/m)/yr, kilogram per meter per year]

Soil texture	Precipitation (mm/yr)			Irrigation (mm/yr)			Runoff (mm/yr)			Recharge (mm/yr)			Evapo-transpiration (mm/yr)			Subsurface drainage (mm/yr)			Change in storage (mm/yr)			Sediment yield [(kg/m)/yr]		
	25	50	75	25	50	75	25	50	75	25	50	75	25	50	75	25	50	75	25	50	75	25	50	75
										Arid														
Clay	140	169	199	1,529	1,529	1,529	658	686	731	136	162	179	813	857	881	0	0	0	−14.3	1.3	13.4	0.05	0.21	0 59
Silty clay	140	169	199	1,529	1,529	1,529	668	703	745	130	149	164	809	854	882	0	0	0	−15.1	0.9	13.9	0.08	0.26	0.70
Silty clay loam	140	169	199	1,529	1,529	1,529	302	325	349	508	538	580	811	851	866	0	0	0	−14.0	−0.9	13.9	0.00	0.13	0 53
Silt	140	169	199	1,529	1,529	1,529	101	112	137	709	759	796	802	847	862	0	0	0	−14.0	−0.7	13.3	0.00	0.00	0 26
Clay loam	140	169	199	1,529	1,529	1,529	139	153	177	664	716	752	806	849	865	0	0	0	−15.3	−1.3	14.7	0.00	0.01	0.08
Silt loam	140	169	199	1,529	1,529	1,529	0	0	1	827	869	926	796	843	858	0	0	0	−13.2	0.3	13.7	0.00	0.00	0.00
Loam	140	169	199	1,529	1,529	1,529	213	235	259	603	642	686	792	840	855	0	0	0	−17.8	−1.3	14.2	0 12	0.25	0.67
Sandy clay loam	140	169	199	1,529	1,529	1,529	186	209	230	638	680	718	781	829	846	0	0	0	−11.0	−0.2	13.4	0.00	0.19	0 90
Sandy clay	140	169	199	1,529	1,529	1,529	103	118	137	728	775	815	777	824	841	0	0	0	−9.9	−0.4	11.0	0.00	0.00	0.08
Loamy sand	140	169	199	1,529	1,529	1,529	0	0	0	842	885	938	779	828	845	0	0	0	−7.0	−0.6	9.8	0.00	0.00	0.00
Sandy loam	140	169	199	1,529	1,529	1,529	0	0	0	841	880	943	781	828	845	0	0	0	−8.7	0.4	11.0	0.00	0.00	0.00
Sand	140	169	199	1,529	1,529	1,529	0	0	0	888	931	982	732	781	801	0	0	0	−7.1	−0.1	9.3	0.00	0.00	0.00
										Humid														
Clay	952	1,064	1,217	0	0	0	213	282	350	0	0	0	722	796	862	0	0	0	−21.9	−4.6	29.2	1.73	3.72	5.79
Silty clay	952	1,064	1,217	0	0	0	187	251	329	0	0	14	747	822	878	0	0	0	−29.4	0.9	28.3	1 39	3.83	5 56
Silty clay loam	952	1,064	1,217	0	0	0	104	150	207	32	79	120	787	839	903	0	0	0	−44.3	9.2	45.4	0.60	2.24	4 59
Silt	952	1,064	1,217	0	0	0	42	80	131	67	147	191	812	853	918	0	0	0	−67.0	11.2	70.3	0 25	1.67	4 34
Clay loam	952	1,064	1,217	0	0	0	78	121	191	92	170	218	747	793	847	0	0	0	−53.9	1.1	59.9	0 51	1.60	3 12
Silt loam	952	1,064	1,217	0	0	0	24	54	97	120	215	272	771	828	878	0	0	0	−83.9	−1.9	75.7	0 32	1.33	3.66
Loam	952	1,064	1,217	0	0	0	53	89	148	152	228	284	705	768	829	0	0	0	−55.2	1.3	59.5	0.66	1.86	3.78
Sandy clay loam	952	1,064	1,217	0	0	0	88	135	205	140	202	266	676	745	801	0	0	0	−47.6	0.9	55.3	1 57	3.72	6.00
Sandy clay	952	1,064	1,217	0	0	0	101	143	216	164	236	289	642	707	766	0	0	0	−43.4	2.1	51.0	1 16	2.45	4 25
Loamy sand	952	1,064	1,217	0	0	0	32	63	90	253	325	384	627	692	750	0	0	0	−35.4	1.0	36.9	0.46	0.98	1 51
Sandy loam	952	1,064	1,217	0	0	0	23	51	75	232	308	359	667	731	794	0	0	0	−47.8	6.7	50.3	0 23	0.63	1 18
Sand	952	1,064	1,217	0	0	0	26	52	82	306	378	449	575	646	706	0	0	0	−28.8	2.5	35.4	0 20	0.41	0.69

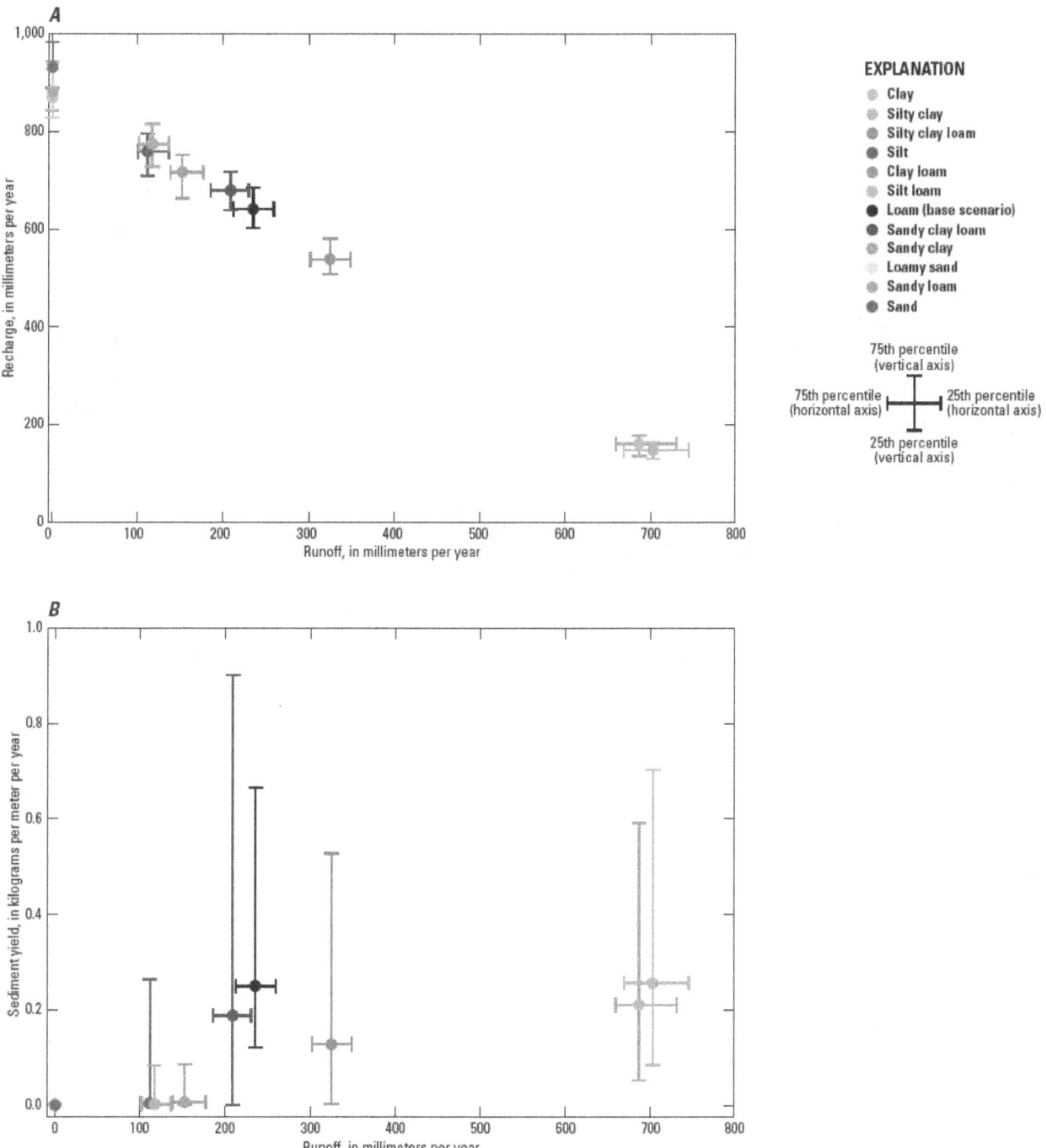

Figure 8.   60-year median values bracketed by the 25th and 75th percentiles (range bars) of relations between (*A*) runoff and recharge and (*B*) runoff and sediment yield as affected by soil texture in an arid climate with sprinkler irrigation.

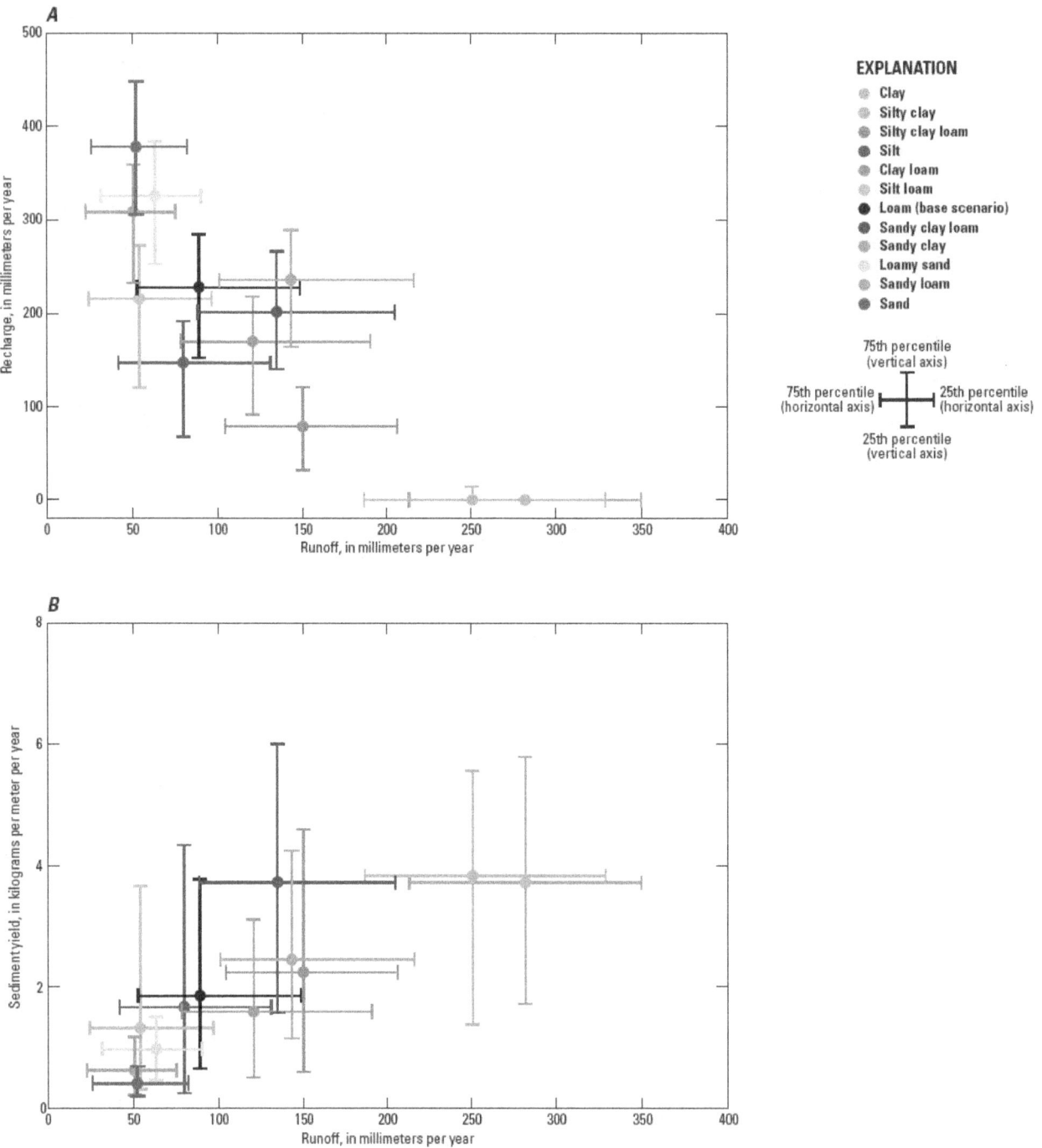

Figure 9.   60-year median values bracketed by the 25th and 75th percentiles (range bars) of relations between (*A*) runoff and recharge and (B) runoff and sediment yield as affected by soil texture in a humid environment.

Effects of Tillage

The timing and type of tillage used on an agricultural field can affect the field-scale water budget and sediment yield (Myers, 1996; Ghidey and Alberts, 1998). Tillage can make the land surface rougher and more (or less) permeable, which affects both runoff and infiltration (Unger, 1992). Conventional moldboard tillage typically breaks up or otherwise disturbs approximately the top 30 cm of soil, which greatly increases the potential for erosion. The timing of tillage relative to seasonal precipitation also can affect the flow paths and sediment yield from a field.

In these simulated scenarios, the type and timing of tillage had a relatively small effect on the partitioning of average annual water budget outflows in both arid and humid environments (tables 6 and 7, figs. 10A and 10B). In the arid environment, there was no clear effect of tillage type on the water budget components. In the humid simulations, recharge was the water budget component most affected by tillage type (fig. 10B). In the arid environment, the no-till option produced the greatest runoff as well as the smallest sediment yield, thereby illustrating the predicted conservation effect of crop residues on the land surface and undisturbed top soil (table 7, fig. 11B). Although the other tillage practices resulted in similar runoff amounts in the arid scenarios, sediment yields varied substantially, and were greatest for reduced and conventional fall tillages. In the humid environment, runoff volume increased as simulated tillage practices were varied from conservation types, such as no till and contour till, to conventional types (table 7, fig. 12A). The association between runoff volume and sediment yield also was very strong for tillage type in the humid environment (fig. 12B). In both the arid and humid environments, conservation tillage practices (no till, conservation till, and contour till) exhibited the lowest sediment yields.

Table 6.   Mean annual values and percentages of overall water budget for each water budget component and sediment yield calculated for a 60-year simulation period as a function of tillage type.

[The arid environment is representative of Sunnyside, Wash., with sprinkler irrigation, and the humid environment is representative of Greenfield, Ind. **Abbreviations**: mm/yr, millimeter per year; (kg/m)/yr, kilogram per meter per year]

Tillage type	Season	Precipitation (mm/yr)	(percent)	Irrigation (mm/yr)	(percent)	Runoff (mm/yr)	(percent)	Recharge (mm/yr)	(percent)	Evapo-transpiration (mm/yr)	(percent)	Subsurface drainage (mm/yr)	(percent)	Change in storage (mm/yr)	(percent)	Sediment yield [(kg/m)/yr]
							Arid									
No till		173	10	1,529	90	287	17	631	37	783	46	0	0	0.3	0.0	0.00
Conservation	Fall	173	10	1,529	90	260	15	636	37	806	47	0	0	−0.1	0.0	0.43
Reduced contour	Fall	173	10	1,529	90	260	15	645	38	797	47	0	0	0.1	0.0	0.01
Reduced	Spring	173	10	1,529	90	261	15	645	38	797	47	0	0	0.1	0.0	0.60
Reduced	Fall	173	10	1,529	90	260	15	622	37	820	48	0	0	−0.2	0.0	0.68
Conventional	Fall	173	10	1,529	90	260	15	620	36	822	48	0	0	−0.1	0.0	0.74
							Humid									
No till		1,088	100	0	0	90	8	229	21	769	71	0	0	0.2	0.0	0.11
Conservation	Fall	1,088	100	0	0	99	9	202	19	787	72	0	0	−0.1	0.0	2.00
Reduced contour	Fall	1,088	100	0	0	82	8	233	21	773	71	0	0	0.1	0.0	0.26
Reduced	Spring	1,088	100	0	0	102	9	222	20	763	70	0	0	0.1	0.0	2.54
Reduced	Fall	1,088	100	0	0	114	11	176	16	798	73	0	0	−0.7	−0.1	2.95
Conventional	Fall	1,088	100	0	0	113	10	175	16	800	73	0	0	−0.6	−0.1	3.35

Table 7. Percentiles (25th, 50th, and 75th) for each water budget component and sediment yield calculated for a 60-year simulation period as a function of tillage type.

[The arid environment is representative of Sunnyside, Wash., with sprinkler irrigation, and the humid environment is representative of Greenfield, Ind. **Abbreviations:** mm/yr, millimeter per year; (kg/m)/yr, kilogram per meter per year]

Tillage type	Season	Precipitation (mm/yr)			Irrigation (mm/yr)			Runoff (mm/yr)			Recharge (mm/yr)			Evapo-transpiration (mm/yr)			Subsurface drainage (mm/yr)			Change in storage (mm/yr)			Sediment yield [(kg/m)/yr]		
		25	50	75	25	50	75	25	50	75	25	50	75	25	50	75	25	50	75	25	50	75	25	50	75
Arid																									
No–till		140	169	199	1,529	1,529	1,529	246	268	292	591	632	673	772	811	839	0	0	0	−15.9	−0.6	12.6	0.00	0.00	0.00
Conservation	Fall	140	169	199	1,529	1,529	1,529	214	236	259	590	627	678	805	850	867	0	0	0	−15.6	−1.1	12.1	0.05	0.12	0.43
Reduced contour	Fall	140	169	199	1,529	1,529	1,529	213	235	259	603	642	686	792	840	855	0	0	0	−18.0	−1.1	14.2	0.00	0.00	0.01
Reduced	Spring	140	169	199	1,529	1,529	1,529	213	235	259	603	642	686	792	840	855	0	0	0	−17.8	−1.3	14.2	0.12	0.25	0.67
Reduced	Fall	140	169	199	1,529	1,529	1,529	213	235	258	582	608	666	821	862	883	0	0	0	−17.4	−1.6	17.5	0.15	0.32	0.73
Conventional	Fall	140	169	199	1,529	1,529	1,529	213	235	257	581	607	659	821	865	885	0	0	0	−12.9	−1.6	13.8	0.18	0.36	0.78
Humid																									
No–till		952	1,064	1,217	0	0	0	43	82	122	154	240	286	711	771	838	0	0	0	−55.8	−6.1	60.8	0.04	0.09	0.15
Conservation	Fall	952	1,064	1,217	0	0	0	57	83	141	136	205	258	718	790	855	0	0	0	−52.2	6.0	61.6	0.48	1.40	3.14
Reduced contour	Fall	952	1,064	1,217	0	0	0	32	76	119	158	240	292	714	777	839	0	0	0	−57.4	3.5	58.6	0.06	0.19	0.43
Reduced	Spring	952	1,064	1,217	0	0	0	53	89	148	152	228	284	705	768	829	0	0	0	−55.2	1.3	59.5	0.66	1.86	3.78
Reduced	Fall	952	1,064	1,217	0	0	0	67	96	162	114	175	232	734	797	864	0	0	0	−54.0	−1.5	62.7	0.92	2.30	4.63
Conventional	Fall	952	1,064	1,217	0	0	0	61	97	168	117	175	233	734	801	867	0	0	0	−52.4	4.8	60.1	1.10	2.65	5.45

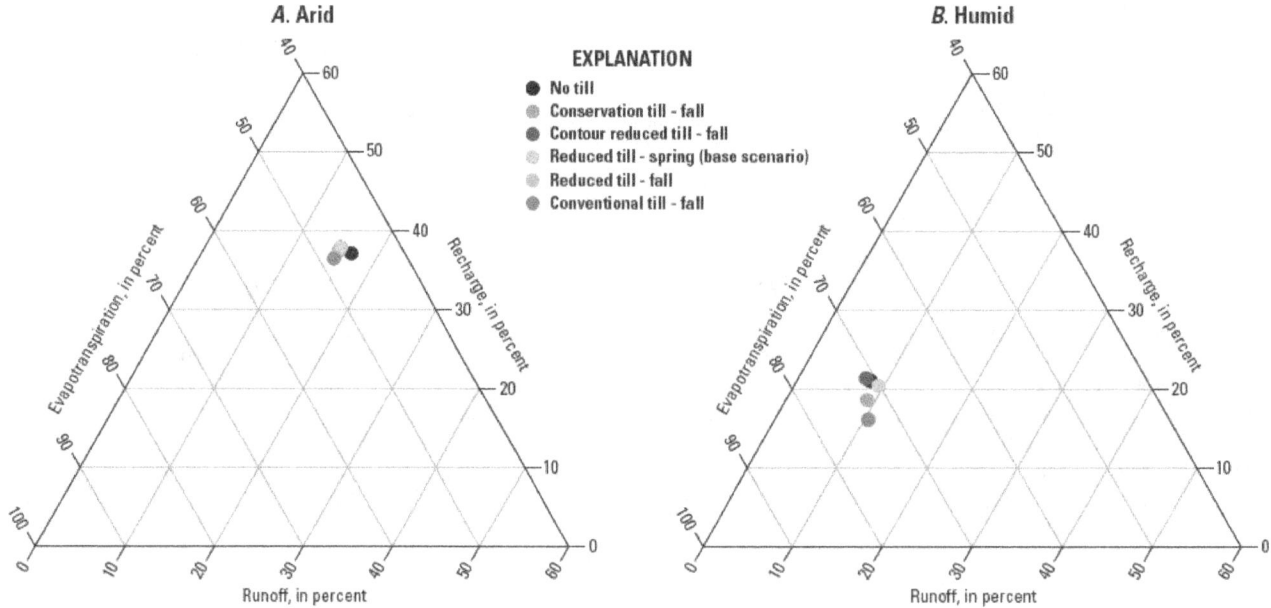

Figure 10. 60-year mean annual percentage of water budget outflows attributable to runoff, evapotranspiration, and recharge as affected by tillage practice for (*A*) an arid environment with sprinkler irrigation and (*B*) a humid environment.

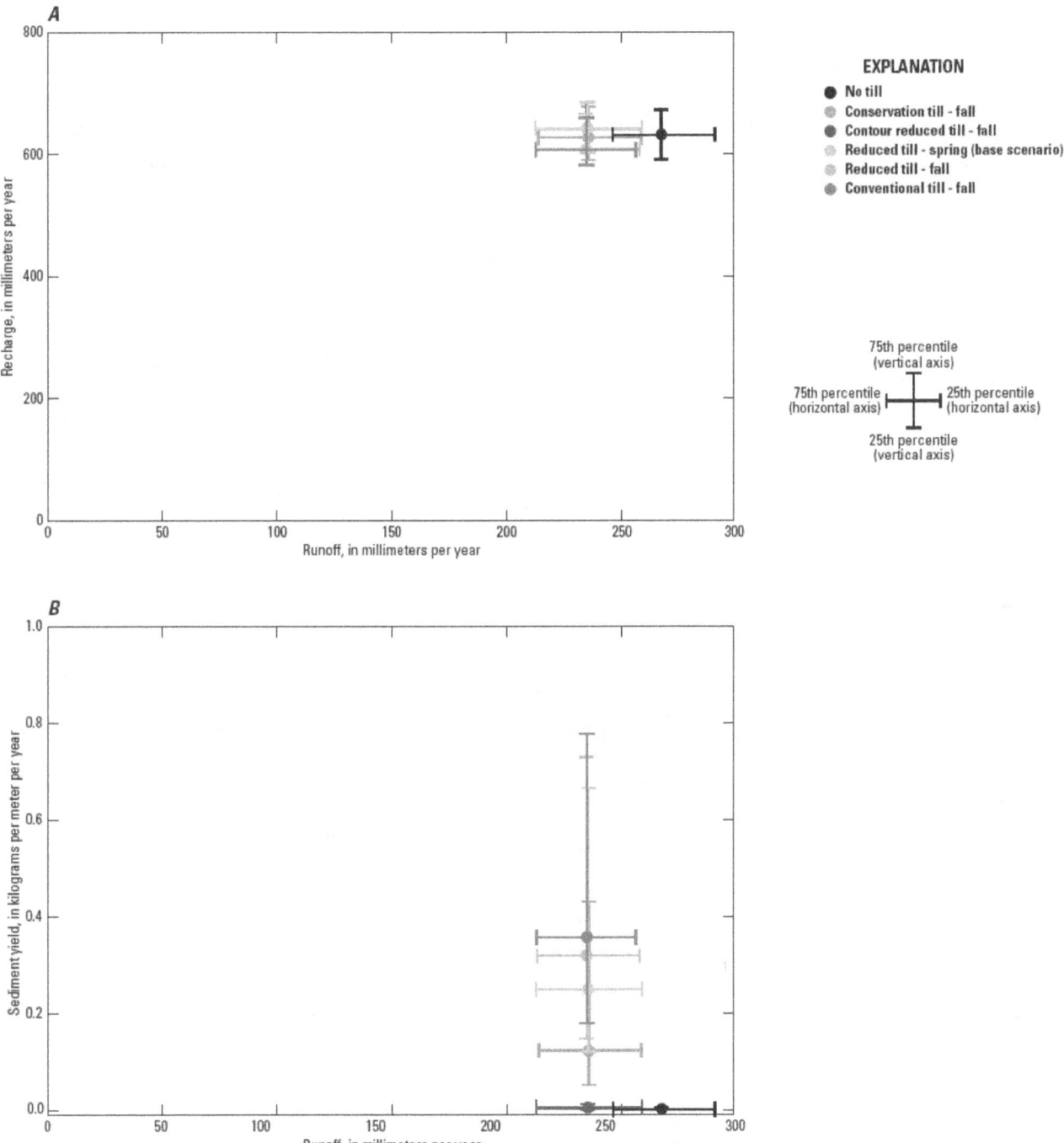

Figure 11. 60-year median values bracketed by the 25th and 75th percentiles (range bars) of relations between (*A*) runoff and recharge and (*B*) runoff and sediment yield as affected by tillage type in an arid environment with sprinkler irrigation.

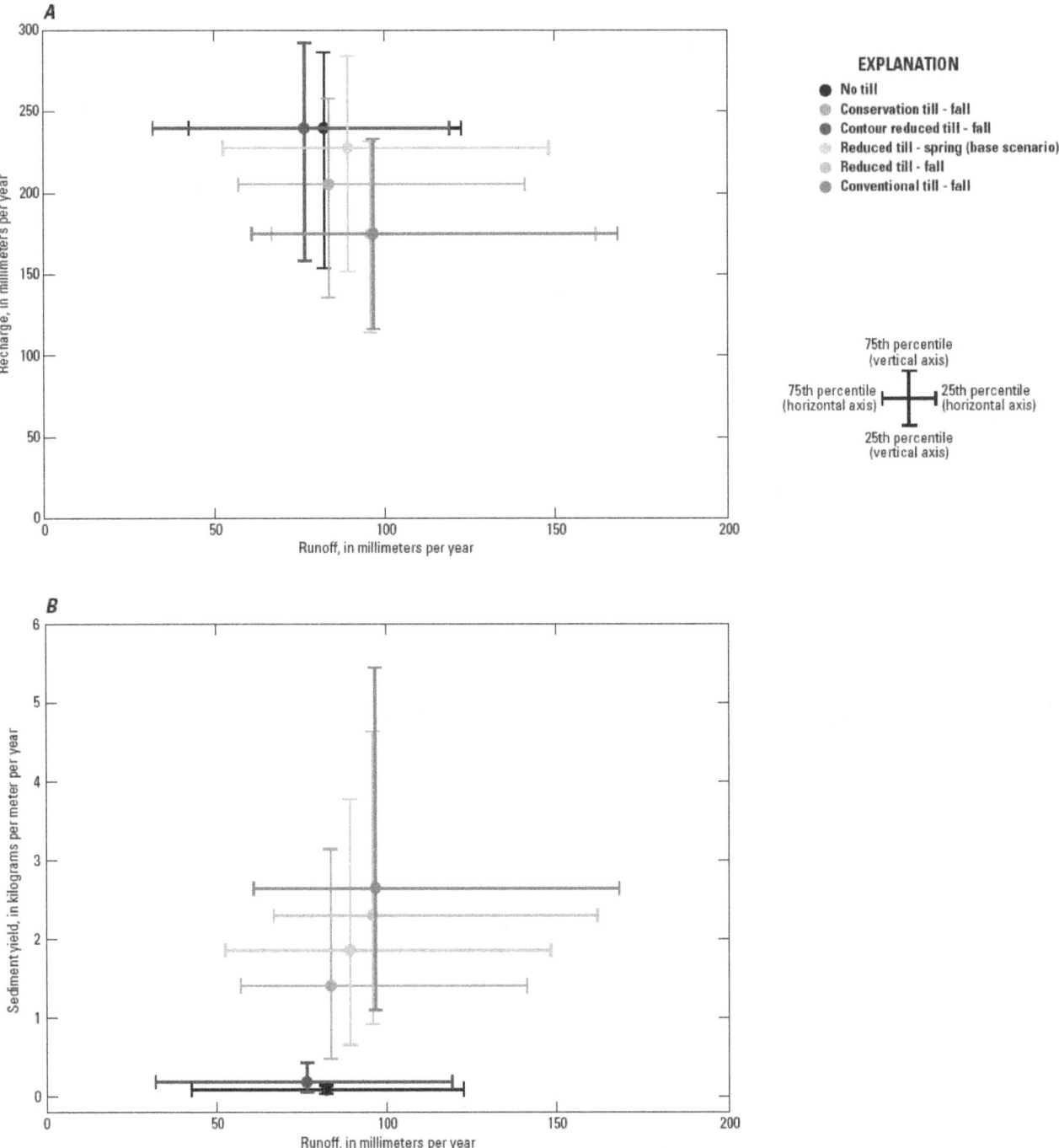

Figure 12.   60-year median values bracketed by the 25th and 75th percentiles (range bars) of relations between (*A*) runoff and recharge and (*B*) runoff and sediment yield as affected by tillage type in a humid environment.

Effects of Land Cover or Crop Type

Land cover, or crop type is a critical determinant of hydrology and sediment transport on the landscape. Different vegetative land covers have different transpiration potentials and plant densities. Thus, different vegetative covers affect water budgets and sediment yield by affecting evapotranspiration, interception of precipitation, and the surface roughness of the landscape (Clark, 1940; van Dijk and Bruijnzeel, 2001). Also coupled with various crop types are the associated agricultural activities, such as the type and timing of tillage, planting, harvest types and timings.

In the arid environment, runoff, as a percentage of total water budget outflow, varied least (of the three outflows) among the various land covers, ranging from 12 to 19 percent (table 8, fig. 13A). Differences in the amount of recharge among the arid scenarios were largest, ranging from 24 to 40 percent of the water budget outflows. Land cover had

a greater effect on the distribution of water budget outflows in the humid environment than in the arid environment (table 8, fig. 13). For the humid environment, runoff ranged from 6 percent for a prairie grass to 20 percent for fallow. Recharge ranged from 2 percent for forest to 39 percent for fallow tilled. Evapotranspiration accounted for just 42 percent of outflow in the untilled fallow scenario, but for 87 percent of outflow in the forested scenario. Annual differences in runoff volumes are much less than annual differences in recharge for the arid scenarios (table 9, fig. 14A). Although the differences in runoff for the arid scenarios are minimal, the differences in sediment yield among different land cover scenarios are substantial. The annual variability in sediment yield from the fallow scenarios was much greater than from scenarios with crops or natural vegetation (fig. 14B). The variability in annual runoff and recharge were much larger in the humid environment than in the arid environment (table 9, fig. 15A), and sediment yields generally were larger in the humid environment.

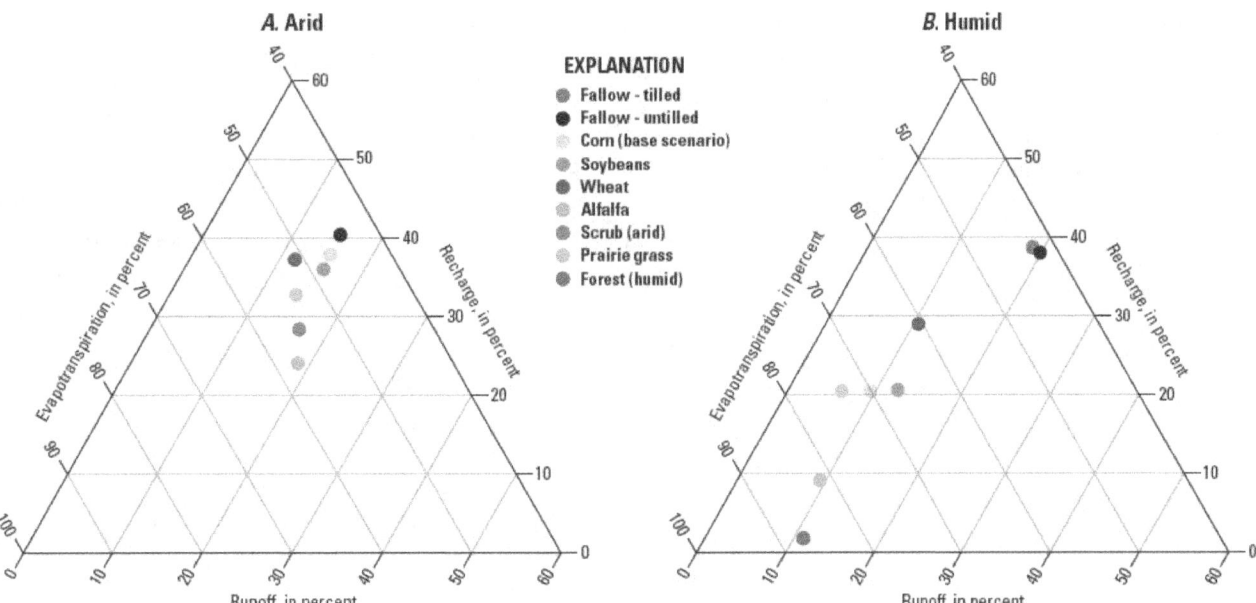

Figure 13. 60-year mean annual percentage of water budget outflows attributable to runoff, evapotranspiration, and recharge as affected by various land covers and crop types for (A) an arid environment with sprinkler irrigation and (B) a humid environment.

Table 8.   Mean annual values and percentages of overall water budget for each water budget component and sediment yield as a function of various land covers and crop types calculated for a 60-year simulation period.

[The arid environment is representative of Sunnyside, Wash., with sprinkler irrigation, and the humid environment is representative of Greenfield, Ind. **Abbreviations**: mm/yr, millimeter per year; (kg/m)/yr, kilogram per meter per year]

Land cover/ crop	Precipitation		Irrigation		Runoff		Recharge		Evapo-transpiration		Subsurface drainage		Change in storage		Sediment yield
	(mm/yr)	(percent)	(mm/yr)	(percent)	(mm/yr)	(percent)	(mm/yr)	(percent)	(mm/yr)	(percent)	(mm/yr)	(percent)	(mm/yr)	(percent)	[(kg/m)/yr]
Arid															
Fallow	173	10	1,529	90	258	15	688	40	757	44	0	0	−0.2	0.0	0.92
Fallow (untilled)	173	10	1,529	90	258	15	687	40	757	44	0	0	−0.3	0.0	0.26
Corn	173	10	1,529	90	261	15	645	38	797	47	0	0	0.1	0.0	0.60
Soybeans	173	10	1,529	90	264	15	614	36	825	48	0	0	−0.3	0.0	0.90
Wheat[1]	173	13	1,165	87	157	12	499	37	683	51	0	0	−0.2	0.0	0.02
Alfalfa[1]	173	11	1,370	89	287	19	372	24	884	57	0	0	−0.3	0.0	0.01
Scrub irrigated	173	10	1,529	90	283	17	483	28	936	55	0	0	−0.2	0.0	0.01
Prairie grass irrigated	173	10	1,529	90	239	14	558	33	905	53	0	0	−0.1	0.0	0.00
Humid															
Fallow	1,088	100	0	0	203	19	421	39	463	43	0	0	0.8	0.1	9.31
Fallow (untilled)	1,088	100	0	0	216	20	413	38	458	42	0	0	0.8	0.1	4.74
Corn	1,088	100	0	0	102	9	222	20	763	70	0	0	0.1	0.0	2.54
Soybeans	1,088	100	0	0	135	12	224	21	729	67	0	0	0.0	0.0	5.27
Wheat	1,088	100	0	0	115	11	315	29	657	60	0	0	0.7	0.1	1.83
Alfalfa	1,088	100	0	0	101	9	100	9	888	82	0	0	−1 1	−0.1	0.33
Forest	1,088	100	0	0	121	11	20	2	949	87	0	0	−1.2	−0.1	0.09
Prairie grass	1,088	100	0	0	67	6	223	20	798	73	0	0	0.8	0.1	0.04

[1]Irrigation timings and amounts differ for both wheat and alfalfa relative to other scenarios due to the time of planting and length of the growing season for these crops.

Table 9. Percentiles (25th, 50th, and 75th) for each water budget component and sediment yield as a function of various land covers and crop types calculated for a 60-year simulation period.

[The arid environment is representative of Sunnyside, Wash., with sprinkler irrigation, and the humid environment is representative of Greenfield, Ind. **Abbreviations**: mm/yr, millimeter per year; (kg/m)/yr, kilogram per meter per year]

Land cover/ crop	Precipitation (mm/yr)			Irrigation (mm/yr)			Runoff (mm/yr)			Recharge (mm/yr)			Evapo-transpiration (mm/yr)			Subsurface drainage (mm/yr)			Change in storage (mm/yr)			Sediment yield [(kg/m)/yr]		
	25	50	75	25	50	75	25	50	75	25	50	75	25	50	75	25	50	75	25	50	75	25	50	75
Arid																								
Fallow	140	169	199	1,529	1,529	1,529	204	232	257	654	678	722	756	796	818	0	0	0	−16 5	−3.9	14 9	0.09	0.56	1.40
Fallow (untilled)	140	169	199	1,529	1,529	1,529	204	233	257	654	678	724	756	795	817	0	0	0	−16 5	−3.9	15.0	0.02	0.09	0.41
Corn	140	169	199	1,529	1,529	1,529	213	235	259	602	642	686	792	840	854	0	0	0	−17.6	−0.5	14 2	0.12	0.25	0.67
Soybeans	140	169	199	1,529	1,529	1,529	217	242	265	570	604	652	825	871	890	0	0	0	−17.4	−1.6	17.6	0.24	0.57	1.28
Wheat[1]	140	169	199	1,165	1,165	1,165	119	144	169	478	491	513	659	698	727	0	0	0	−20.4	1.1	16.4	0.00	0.00	0.01
Alfalfa[1]	140	169	199	1,376	1,376	1,376	244	273	313	317	367	413	828	919	951	0	0	0	−20.0	2.0	25.8	0.01	0.01	0.01
Irrigated scrub	140	169	199	1,529	1,529	1,529	238	266	283	442	469	507	917	966	1,007	0	0	0	−28.6	−3.5	27.8	0.00	0.00	0.01
Irrigated prairie grass	140	169	199	1,529	1,529	1,529	187	215	240	508	558	596	897	928	972	0	0	0	−34 2	−5.0	32.0	0.00	0.00	0.00
Humid																								
Fallow	952	1,064	1,217	0	0	0	134	175	268	376	423	462	421	474	504	0	0	0	−30 2	−3.2	29.7	4.88	8.96	13.39
Fallow (untilled)	952	1,064	1,217	0	0	0	150	188	281	368	418	456	418	466	499	0	0	0	−29.8	−3.0	29 2	2.52	3.91	7.02
Corn	952	1,064	1,217	0	0	0	53	89	148	152	228	284	705	768	829	0	0	0	−55 2	1.3	59 5	0.66	1.86	3.78
Soybeans	952	1,064	1,217	0	0	0	87	121	183	166	225	279	669	746	787	0	0	0	−36.0	3.2	43.0	1.91	4.56	7.50
Wheat	952	1,064	1,217	0	0	0	72	104	143	258	315	383	595	661	730	0	0	0	−28 2	−3.3	26.4	0.39	1.10	2.65
Alfalfa	952	1,064	1,217	0	0	0	63	94	134	34	84	162	799	884	980	0	0	0	−58 1	−5.2	57 2	0.03	0.09	0.16
Forest	952	1,064	1,217	0	0	0	77	116	159	0	0	27	827	958	1,057	0	0	0	−51 3	0.2	57 9	0.02	0.08	0.12
Prairie grass	952	1,064	1,217	0	0	0	30	61	89	162	228	282	713	813	904	0	0	0	−41.4	−1.4	37.7	0.01	0.03	0.07

[1]Irrigation timings and amounts differ for both wheat and alfalfa relative to other scenarios due to the time of planing and length of the growing season for these crops.

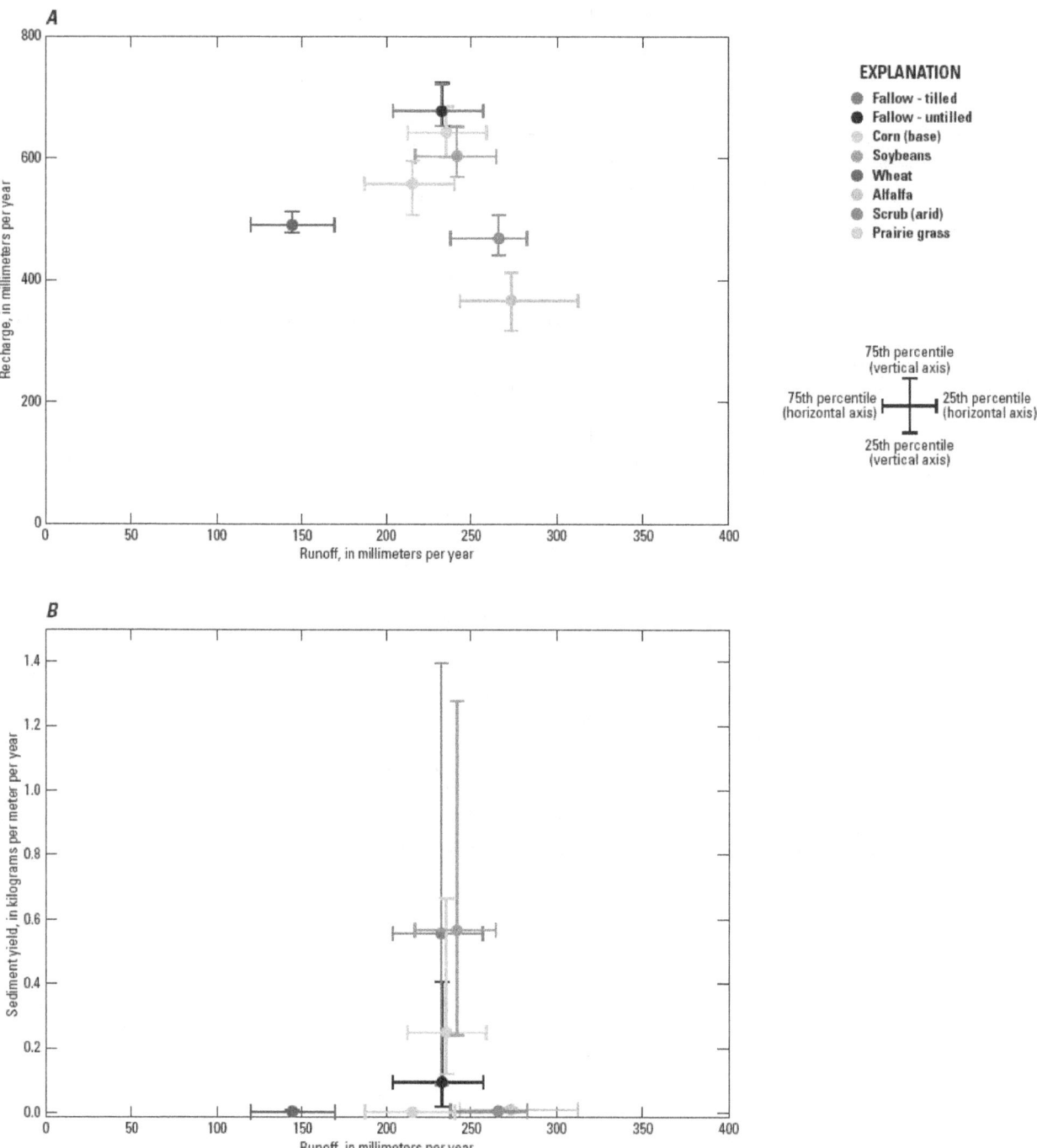

Figure 14. 60-year median values bracketed by the 25th and 75th percentiles (range bars) of relations between (*A*) runoff and recharge and (*B*) runoff and sediment yield as affected by various land covers and crop types in an arid environment with sprinkler irrigation.

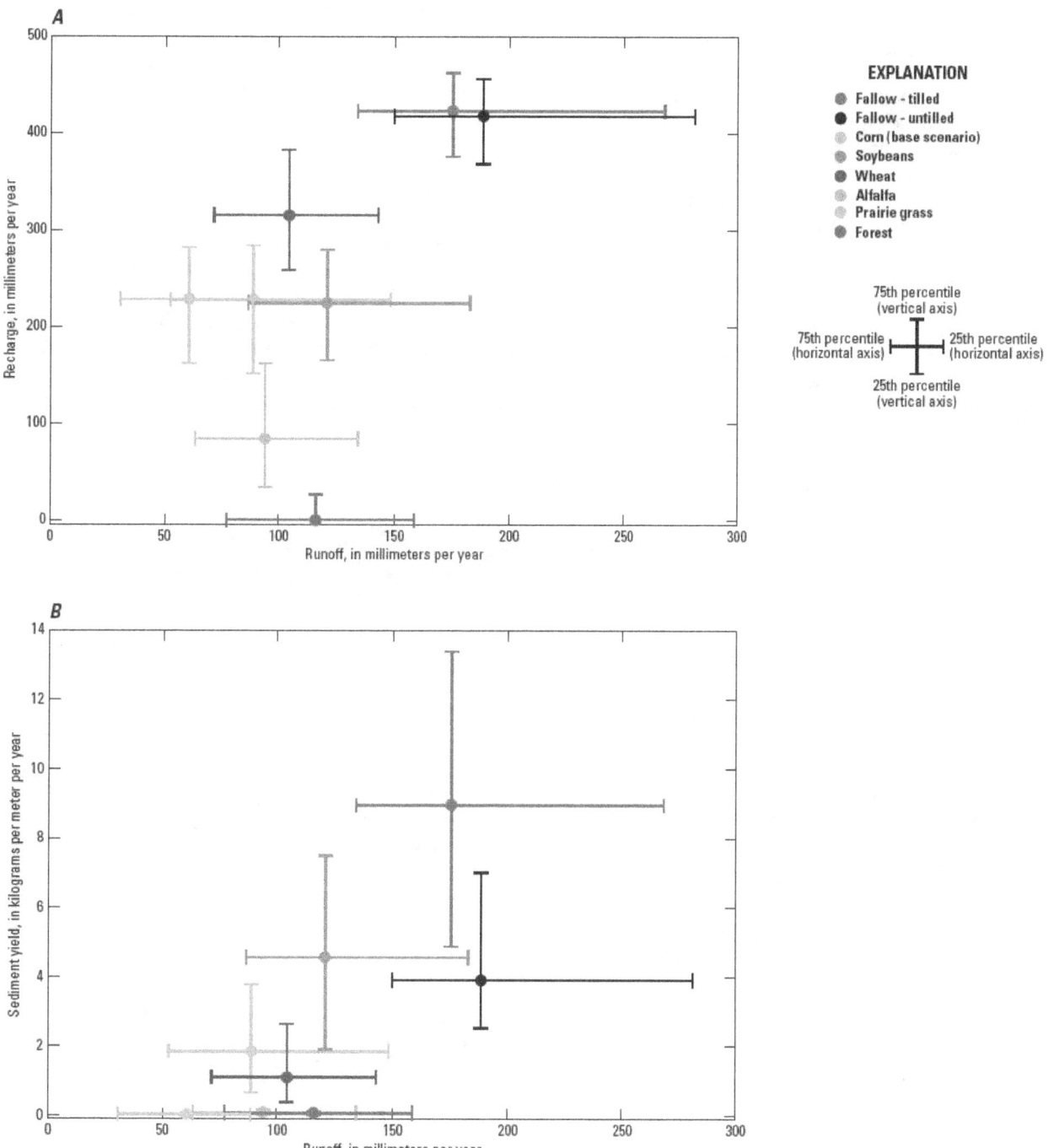

Figure 15. 60-year median values bracketed by the 25th and 75th percentiles (range bars) of relations between (*A*) runoff and recharge and (*B*) runoff and sediment yield as affected by various land covers and crop types in a humid environment.

Effects of Selected Agricultural Land Management Practices

Many common agricultural land management practices (AMPs) affect hydrology and sediment transport at the field scale. Crop rotation, cover cropping, and strip cropping are annually implemented AMPs that blend elements of different crop and tillage regimes, whereas terracing and the installation of subsurface drainage are more permanent AMPs that fundamentally alter the landscape to enhance agricultural production).

The relative percentages of water budget outflows (runoff, recharge, and evapotranspiration) are affected by various AMPs in both the arid and humid environments as shown in figure 16. With the exception of strip cropping

and subsurface drainage, these AMPs had little effect on the outflows in the arid environment (table 10, fig. 16A). Greater variability in the water budget outflows was indicated by simulations for the humid environment (table 10, fig. 16B). In the humid environment simulations, runoff accounted for a similar percentage of the total outflow (8–11 percent) for most AMPs (fig. 16B). In the humid environment, the AMPs mostly affected the distribution between evapotranspiration and recharge. In the arid environment with sprinkler irrigation, recharge and runoff were most affected by subsurface drains and strip cropping, whereas sediment yield was most affected by terracing and crop rotation (table 11, fig. 17). Sediment yield generally increased with runoff, with the exception of the terraced field. Sediment In the humid environment, sediment yield was the greatest for the crop rotation simulation (table 11, fig. 18).

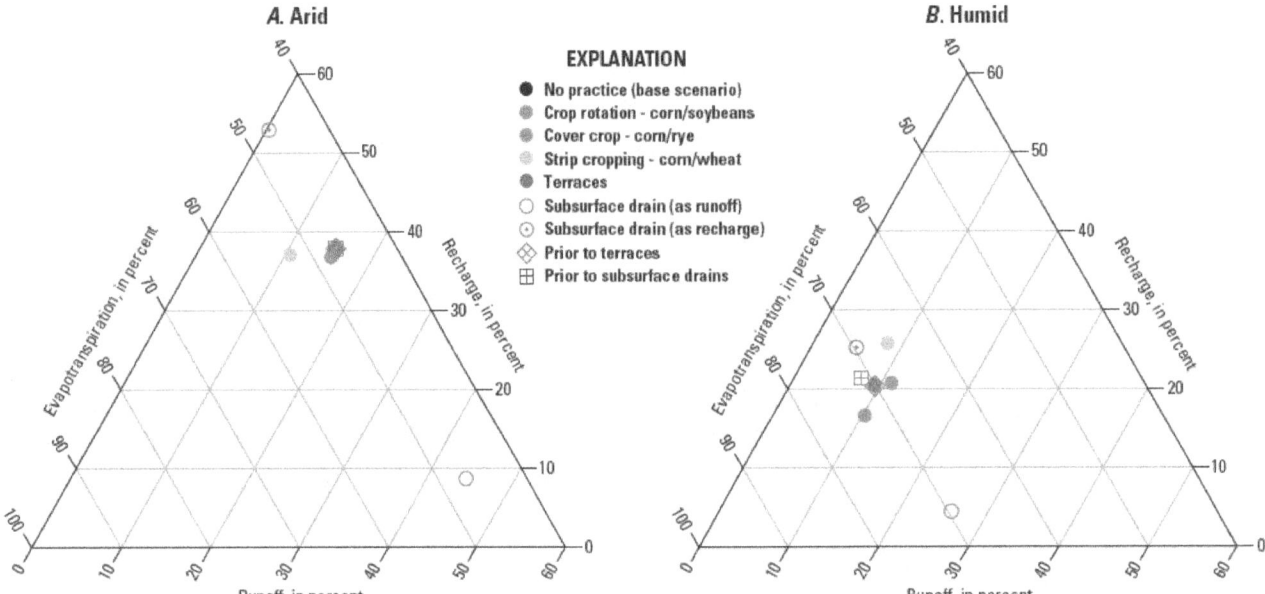

The original subsurface drain data were not included in order to maintain the magnitude of the water budgets for each environment.
The data points for subsurface drain "as runoff" and "as recharge" includes the volume of drainage water as either a portion of runoff or
recharge, respectively. The "prior to terraces" scenario used a 8-percent slope, and the "prior to subsurface drains" scenario used a 0.5-percent slope.

Figure 16.   60-year mean annual percentage of water budget outflows attributable to runoff, evapotranspiration, and recharge as affected by various agricultural land management practices for (*A*) an arid environment with sprinkler irrigation and (*B*) a humid environment.

Table 10. Mean annual values and percentages of overall water budget for each water budget component and sediment yield as a function of various agricultural land management practices calculated for a 60-year simulation period.

[The arid environment is representative of Sunnyside, Wash , with sprinkler irrigation, and the humid environment is representative of Greenfield, Ind. **Abbreviations**: mm/yr, millimeter per year; (kg/m)/yr, kilogram per meter per year; RO, runoff; RC, recharge]

Management/ modification	Precipitation		Irrigation		Runoff		Recharge		Evapo-transpiration		Subsurface drainage		Change in storage		Sediment yield
	(mm/yr)	(percent)	(mm/yr)	(percent)	(mm/yr)	(percent)	(mm/yr)	(percent)	(mm/yr)	(percent)	(mm/yr)	(percent)	(mm/yr)	(percent)	[(kg/m)/yr]
Arid															
None	173	10	1,529	90	261	15	645	38	797	47	0	0	0.1	0.0	0.60
Crop rotation	173	10	1,529	90	262	15	631	37	810	48	0	0	−0.3	0.0	0.91
Cover crop	173	10	1,529	90	260	15	626	37	816	48	0	0	−0.2	0.0	0.68
Strip cropping	173	11	1,347	89	154	11	543	37	767	52	0	0	−0.2	0.0	0.00
Terraces	173	10	1,529	90	261	15	645	38	797	47	0	0	0.1	0.0	0.00
Subsurface drains	173	10	1,529	90	6	1	148	16	796	84	752	44	0.0	0.0	0.00
Subsurface drainage as RO[1]	173	10	1,529	90	758	45	148	9	796	47	0	0	0.0	0.0	0.00
Subsurface drainage as RC[2]	173	10	1,529	90	6	0	900	53	796	47	0	0	0.0	0.0	0.00
Prior to terraces[3]	173	10	1,529	90	261	15	645	38	797	47	0	0	0.1	0.0	0.60
Prior to subsurface drains[4]	173	10	1,529	90	261	15	645	38	797	47	0	0	0.1	0.0	0.01
Humid															
None	1,088	100	0	0	102	9	222	20	763	70	0	0	0.1	0.0	2.54
Crop rotation	1,088	100	0	0	122	11	225	21	741	68	0	0	0.0	0.0	3.87
Cover crop	1,088	100	0	0	112	10	181	17	795	73	0	0	−0.2	0.0	3.21
Strip cropping	1,088	100	0	0	90	8	281	26	720	66	0	0	0.4	0.0	0.27
Terraces	1,088	100	0	0	102	9	222	20	763	70	0	0	0.1	0.0	0.79
Subsurface drains	1,088	100	0	0	55	5	49	5	757	70	227	21	−0.3	0.0	0.20
Subsurface drainage as RO[1]	1,088	100	0	0	281	26	49	5	757	70	0	0	−0.3	0.0	0.20
Subsurface drainage as RC[2]	1,088	100	0	0	55	5	276	25	757	70	0	0	−0.3	0.0	0.20
Prior to terraces[3]	1,088	100	0	0	102	9	222	20	763	70	0	0	0.1	0.0	2.54
Prior to subsurface drains[4]	1,088	100	0	0	81	7	234	21	773	71	0	0	0.1	0.0	0.23

[1]Subsurface drainage volume added to RO volume.

[2]Subsurface drainage volume added to RC volume.

[3]Land slope prior to terracing equal to 8 percent slope.

[4]Land slope prior to subsurface drains equal to 0.5 percent, not 3 percent.

Table 11.   Percentiles (25th, 50th, and 75th) for each water budget component and sediment yield as a function of various agricultural land management practices calculated for a 60-year simulation period.

[The arid environment is representative of Sunnyside, Wash., with sprinkler irrigation, and the humid environment is representative of Greenfield, Ind. **Abbreviations**: mm/yr, millimeter per year; (kg/m)/yr, kilogram per meter per year; RO, runoff; RC, recharge]

Management/ modification	Precipitation (mm/yr)			Irrigation (mm/yr)			Runoff (mm/yr)			Recharge (mm/yr)			Evapo-transpiration (mm/yr)			Subsurface drainage (mm/yr)			Change in storage (mm/yr)			Sediment yield [(kg/m)/yr]		
	25	50	75	25	50	75	25	50	75	25	50	75	25	50	75	25	50	75	25	50	75	25	50	75
Arid																								
None	140	169	199	1,529	1,529	1,529	213	235	259	603	642	686	792	840	855	0	0	0	−17.8	−1.3	14.2	0.12	0.25	0.67
Crop rotation	140	169	199	1,529	1,529	1,529	217	242	259	581	625	681	807	854	874	0	0	0	−37.9	1.6	34.5	0.09	0.53	1.28
Cover Crop	140	169	199	1,529	1,529	1,529	212	235	257	584	610	668	819	860	880	0	0	0	−19.2	−0.4	16.7	0.15	0.33	0.72
Strip cropping[5]	140	169	199	1,347	1,347	1,347	117	141	167	511	538	584	765	797	816	0	0	0	−18.6	−0.9	14.8	0.00	0.00	0.00
Terraces	140	169	199	1,529	1,529	1,529	213	235	259	603	642	686	792	840	855	0	0	0	−17.8	−1.3	14.2	2.40	2.84	3.14
Subsurface drains	140	169	199	1,529	1,529	1,529	0	0	0	124	141	169	792	839	855	700	725	775	−14.7	−0.4	15.1	0.00	0.00	0.00
Subsurface drainage as RO[1]	140	169	199	1,529	1,529	1,529	700	725	775	124	141	169	792	839	855	0	0	0	−14.7	−0.4	15.1	0.00	0.00	0.00
Subsurface drainage as RC[2]	140	169	199	1,529	1,529	1,529	0	0	0	824	866	944	792	839	855	0	0	0	−14.7	−0.4	15.1	0.00	0.00	0.00
Prior to terraces[3]	140	169	199	1,529	1,529	1,529	213	235	259	603	642	686	792	840	855	0	0	0	−17.8	−1.3	14.2	0.12	0.25	0.67
Prior to subsurface drains[4]	140	169	199	1,529	1,529	1,529	213	236	259	603	642	686	792	840	855	0	0	0	−17.8	−1.3	14.2	0.00	0.01	0.01
Humid																								
None	952	1,064	1,217	0	0	0	53	89	148	152	228	284	705	768	829	0	0	0	−55.2	1.3	59.5	0.66	1.86	3.78
Strip cropping[5]	952	1,064	1,217	0	0	0	52	82	120	208	288	339	655	725	793	0	0	0	−32.5	1.6	35.8	0.08	0.20	0.37
Cover crop	952	1,064	1,217	0	0	0	67	92	157	115	182	234	733	796	859	0	0	0	−54.0	5.3	59.4	0.89	2.48	4.81
Crop rotation	952	1,064	1,217	0	0	0	70	110	176	167	217	277	697	745	799	0	0	0	−39.2	−8.4	45.6	1.25	3.23	6.04
Terraces	952	1,064	1,217	0	0	0	53	89	148	152	228	284	705	768	829	0	0	0	−55.2	1.6	59.5	0.28	0.71	1.18
Subsurface drains	952	1,064	1,217	0	0	0	17	46	85	28	50	71	697	758	816	174	208	283	−27.8	−2.7	30.6	0.04	0.13	0.31
Subsurface drainage as RO[1]	952	1,064	1,217	0	0	0	191	254	368	28	50	71	697	758	816	0	0	0	−27.8	−2.7	30.6	0.04	0.13	0.31
Subsurface drainage as RC[2]	952	1,064	1,217	0	0	0	17	46	85	202	257	355	697	758	816	0	0	0	−27.8	−2.7	30.6	0.04	0.13	0.31
Prior to terraces[3]	952	1,064	1,217	0	0	0	53	89	148	152	228	284	705	768	829	0	0	0	−55.2	1.3	59.5	0.66	1.86	3.78
Prior to subsurface drains[4]	952	1,064	1,217	0	0	0	31	74	116	159	240	293	715	778	840	0	0	0	−57.4	3.7	58.6	0.05	0.17	0.39

[1] Subsurface drainage volume added to runoff volume

[2] Subsurface drainage volume added to recharge volume

[3] Land slope prior to terracing equal to 8 percent slope

[4] Land slope prior to subsurface drains equal to 0.5 percent, not 3 percent

[5] Irrigation timings and amounts differ for corn and wheat in strip cropping scenario due to the time of planting and length of the growing season for these crops.

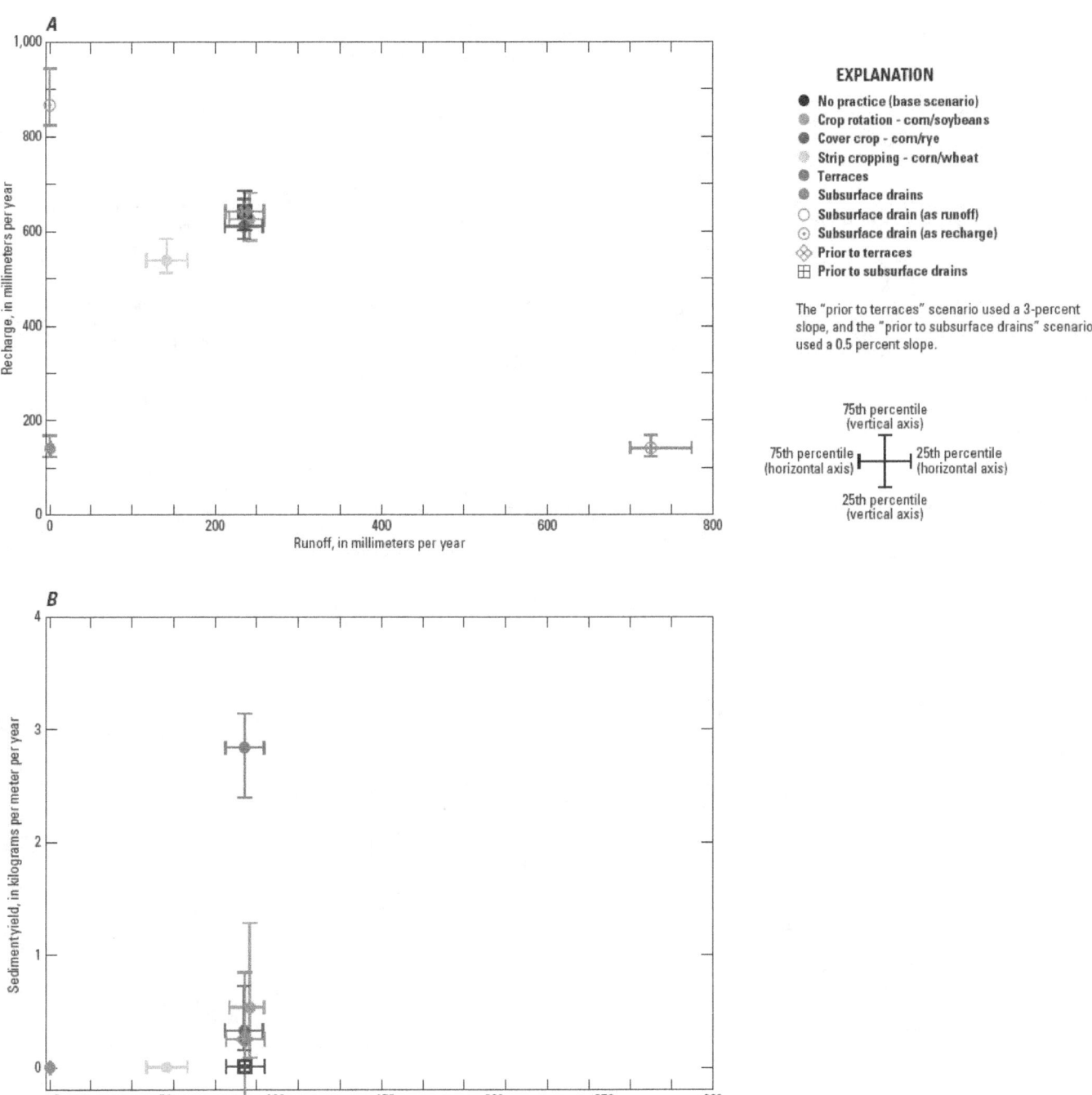

Figure 17.   60-year median values bracketed by the 25th and 75th percentiles (range bars) of relations between (*A*) runoff and recharge and (*B*) runoff and sediment yield as affected by various agricultural management practices in an arid environment with sprinkler irrigation.

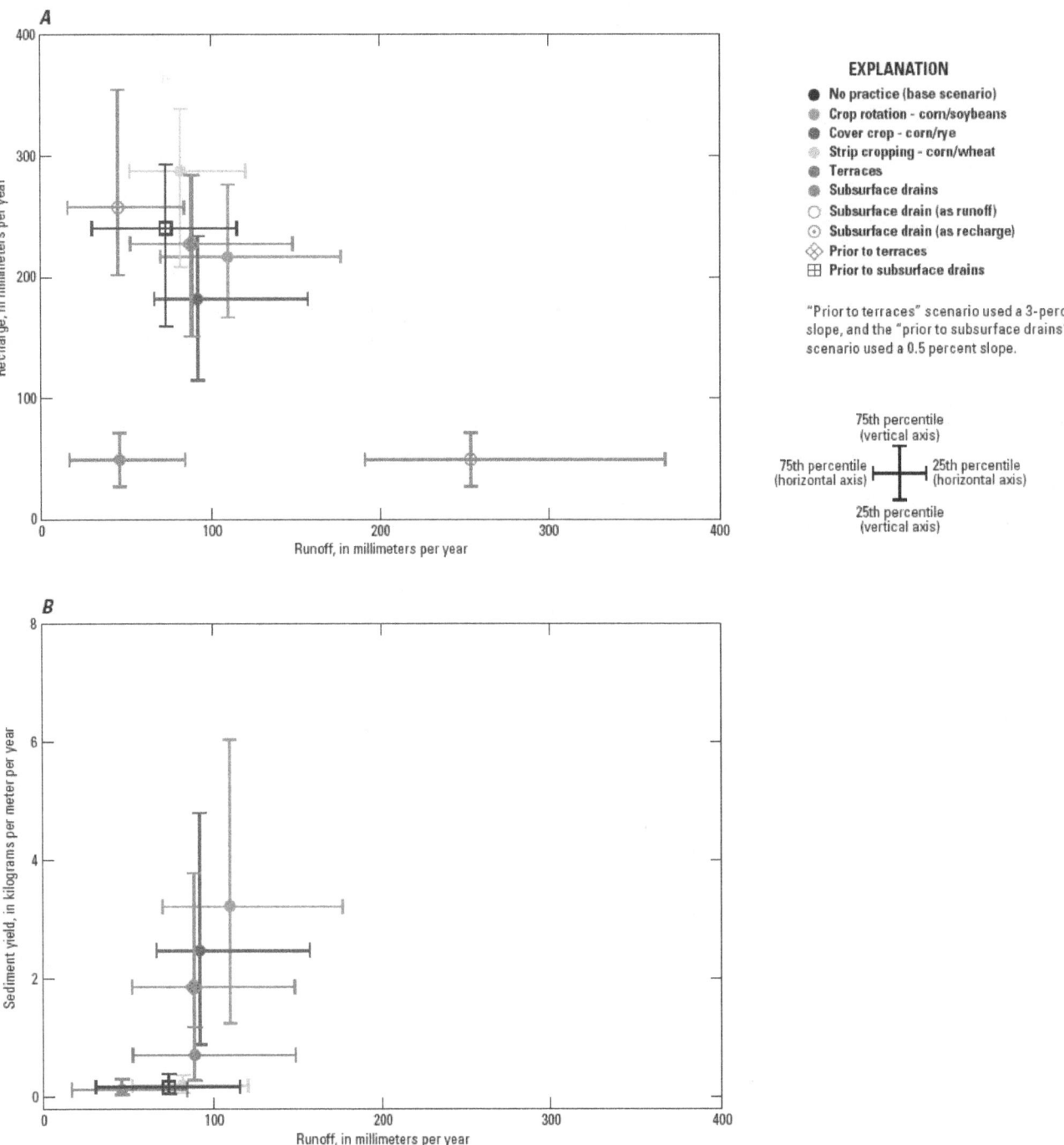

Figure 18.   60-year median values bracketed by the 25th and 75th percentiles (range bars) of relations between (*A*) runoff and recharge and (*B*) runoff and sediment yield as affected by various agricultural management practices in a humid environment.

Effects of Selected Agricultural Land Management Practices Relative to the Base Scenarios

The previous sections of this report described the variations in water budget outflow components and sediment yields as a function of the change of a single landscape characteristic or agricultural management practice characteristic. In this section, the relative differences in water budget outflows and sediment yields for selected agricultural management practices and pre-agricultural scenarios are presented relative to the base arid (fig. 19) and base humid (fig. 20) environment scenarios.

As a percent difference relative to the base scenarios, evapotranspiration is the least affected water budget component among the various simulated agricultural scenarios for both the arid and humid environments. In the arid environment, alfalfa, prairie grass, and scrub had the greatest relative increase for evapotranspiration, whereas, in the humid environment, forest and alfalfa covers had the largest

positive changes in evapotranspiration relative to the base scenario. Generally, in the scenarios for which large increases in evapotranspiration were observed, a corresponding large decrease in recharge was observed. In both climate regimes, changes in runoff and sediment yield generally were negative with respect to the base scenarios, with a few exceptions such as in the soybean and fallow scenarios. In the arid environment, the results of several simulations—for no till, alfalfa, fallow, scrub, and cover crop—showed deviations of runoff and sediment yield in different directions relative the base scenario. In the humid environment, runoff and sediment yield deviated in the same direction from the base scenario for all except one simulation: wheat (for which runoff was greater than for the base scenario but sediment yield was less).

These results, although insightful on their own, may spur further analyses for better understanding of the fundamental processes underlying the hydrologic cycle and the process of soil erosion. Thus, the WEPP model parameters used to generate these data accompanied by all simulation configurations as well as the daily and annualized results presented in this report are provided as appendixes.

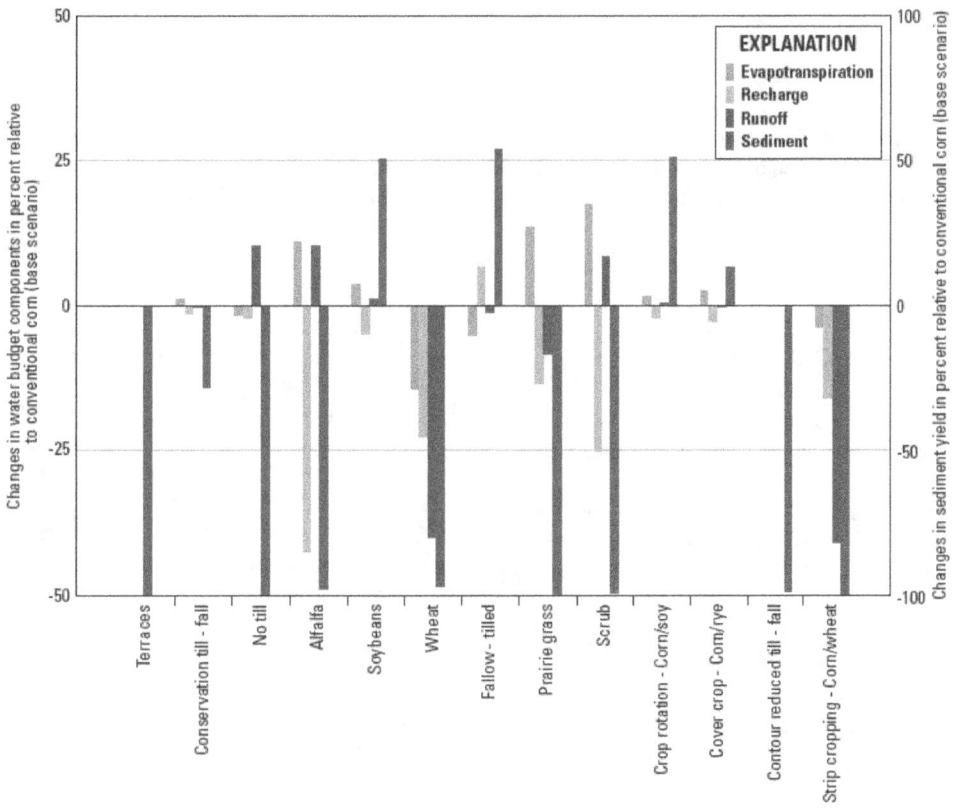

Figure 19. Relative effects, in terms of percent change, of selected agricultural practices and pre-agricultural scenarios on water budget components and sediment yield in an arid environment compared to 60-year annual mean values for the "base" arid environment scenario.

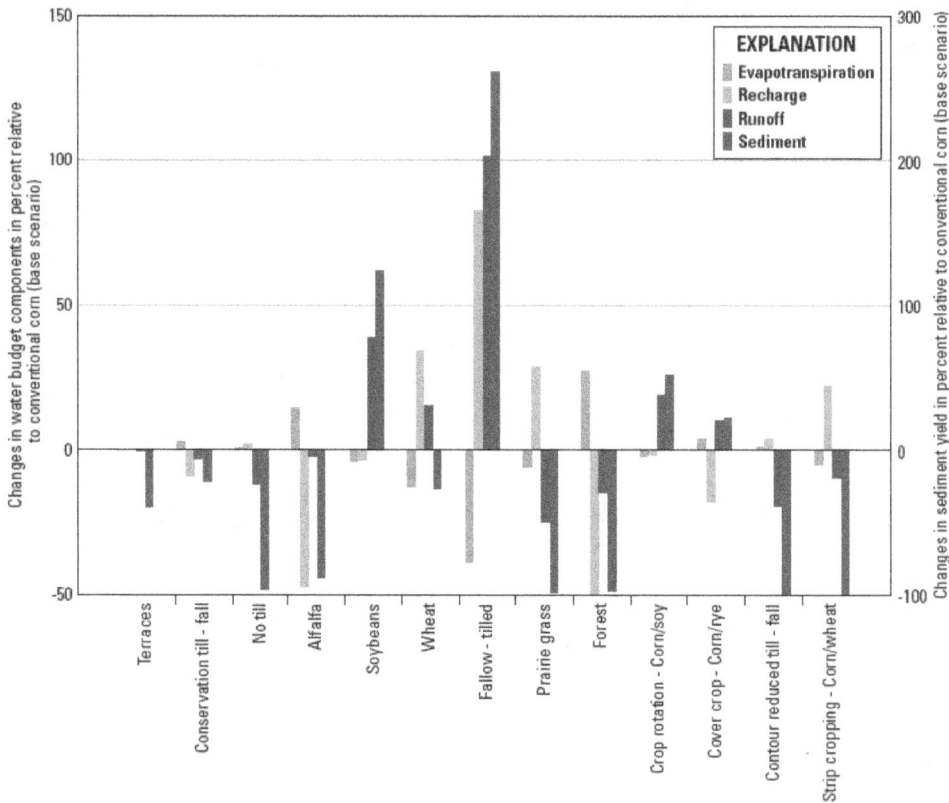

Figure 20.   Relative effects, in terms of percent change, of selected agricultural practices and pre-agricultural scenarios on water budget components and sediment yield in a humid environment compared to 60-year mean values for the "base" humid environment scenario.

Summary and Conclusions

Crop agriculture occupies 13 percent of the conterminous United States. The characteristics of the land that agriculture is practiced upon, such as slope and soil texture, as well as the agricultural management practices implemented on that land, such as crop type and tillage type, have effects on the water budgets of these lands. These hydrologic changes may include changes to the magnitude and partitioning of water budgets and flow paths as well as sediment yields of agricultural lands.

For this report, an array of 68 simulations were performed using the Water Erosion Prediction Project (WEPP) model in which the variables of slope, soil texture, land cover or crop type, tillage type, and selected agricultural management practices were varied to predict potential changes in the water budget and sediment yield of a hypothetical 16.2-hectare agricultural field. Each simulation represented

one of two common agricultural climate regimes: arid with sprinkler irrigation and humid. The results of these simulations demonstrate the magnitudes of potential changes in water budgets and sediment yields from lands with various slopes, soil textures, and agricultural management practices.

In general, less variability in the water budget outflows for simulations in the arid environment was predicted than for the equivalent scenarios in the humid environment. Also, interannual variability within the 60-year simulations generally was less in the arid environment scenarios when compared to the equivalent humid environment scenarios. The lesser variability in water budget outflows within the arid scenarios may be explained by the steady uniform delivery of irrigation water simulated within the arid scenarios. By comparison, the natural variability of the occurrence, intensity, and amount of precipitation within the humid environment resulted in the greater variability amongst water budgets and sediment yields.

Land slope is a landscape characteristic that is often a determinant of the amount of runoff and erosion occurring from an agricultural field. In the arid climate scenarios, little change in water budget outflows were observed as slope ranged from 0.5 to 10 percent. Sediment yield in the arid simulations, however, increased linearly with increases in land slope despite negligible observed changes in runoff. In the humid climate scenarios, runoff increased with land slope while evapotranspiration and recharge decreased. Sediment yield also increased linearly with increased land slope in the humid environment. In both environments, the scenarios simulating a 10-percent land slope exhibited the greatest sediment yields.

The effect of soil texture on water budgets and sediment yields also were investigated. Soil texture may affect the movement of water through soil as well as the erodibility of a soil. In general for both the arid and humid climates, soil texture scenario groups, runoff and sediment yield increased from clayey to sandy textured soils, whereas recharge and evapotranspiration tended to decrease. In both environments, the scenarios for clay and silty clay soils generated more runoff than any of the other soil textures.

Agricultural practices concerning the timing and type of tillage used on an agricultural field may affect the water budget and sediment yield of the land by altering the roughness and permeability of the land surface and turning over vegetation and residue. In the tillage scenarios, the type and timing of tillage had little effect on the mean annual water budget outflows for both the arid and humid environments. Although the amount of water in runoff was affected little by tillage practices, the different types of tillage practices have a stronger affect on sediment yields in both environments. Conservation tillage practices, such as no till, conservation till, and contour till, yielded less sediment than conventional and reduced tillage.

The land cover or crop type that is present on a field can directly affect every outflow component of a water budget. Vegetative cover intercepts and transpires precipitation, while roots wick infiltrated water from the unsaturated zone reducing recharge. Also, the density and roughness of vegetation can impede the runoff of excess precipitation. In the land cover and crop type scenarios for both the arid and humid environments, evapotranspiration increased over the spectrum of land covers from fallow to agricultural crops and then to the pre-agricultural land covers, while recharge and runoff both decreased over the same spectrum. Sediment yields within both environments were greatest for the fallow and soybean scenarios and least for the alfalfa, wheat, and natural land covers.

Agricultural land management practices (AMPs) are implemented for a variety of purposes; some of which may include altering field-scale hydrology and reducing sediment transport. In the arid environment, the AMPs simulated for this report (strip cropping, cover crop, crop rotation, terracing, subsurface drainage) had minor effects on partitioning of water budget outflows, with the exception of strip cropping and subsurface drainage. Simulation of AMPs in the humid environment showed greater variability in the partitioning of water budget outflows with cover cropping, strip cropping, and subsurface drainage showing the greatest differences. The strip cropping and subsurface drainage scenarios exhibited the lowest sediment yields for both environments while annual crop rotation of corn and soybeans increased the sediment yield compared to the base scenario continuous corn.

Acknowledgments

Technical support for the Water Erosion Prediction Project (WEPP) model was provided by Jim Frankenberger of the U.S. Department of Agriculture's National Erosion Research Laboratory (NSERL). Technical reviews were provided by Erick Burns and David Lampe, U.S. Geological Survey.

References Cited

Agarwal, A., and Dickinson, W.T., 1991, Effect of texture, flow, and slope on interrill sediment transport: Transactions of the American Society of Agricultural Engineers, v. 34, p. 1,726–1,731.

Baker, N.T., and Capel, P.D., 2011, Environmental factors that influence the location of crop agriculture in the conterminous United States: U.S. Geological Survey Scientific Investigations Report 2011–5108, 72 p.

Böhlke, J.K., 2002, Groundwater recharge and agricultural contamination: Hydrogeology Journal, v. 10, no. 1, p. 153–179.

Bosch, J., and Hewlett, J., 1982, A review of catchment experiments to determine the effect of vegetation changes on water yield and evapotranspiration: Journal of Hydrology, v. 55, no. 1-4, p. 3–23.

Brakensiek, D., Engleman, R., and Rawls, W., 1981, Variation within texture classes of soil water parameters: Transactions of the American Society of Agricultural Engineers, v. 24, no. 2, p. 335–339.

Carpenter, S.R., Caraco, N.F., Correll, D.L., Howarth, R.W., Sharpley, A.N., and Smith, V.H., 1998, Nonpoint pollution of surface waters with phosphorus and nitrogen: Ecological Applications, v. 8, no. 3, p. 559–568.

Chow, V.T., Maidment, D.R., and Mays, L.W., 1988, Applied hydrology: New York, McGraw-Hill, 572 p.

Chu, S.T., 1978, Infiltration during an unsteady rain: Water Resources Research, v. 14, no. 3, p. 461.

Claassen, R., Carriazo, F., Cooper, J.C., Hellerstein, D., and Ueda, K., 2011, Grassland to cropland conversion in the Northern Plains—The role of crop insurance: U.S. Department of Agriculture, Economic Research Service, Commodity and Disaster Programs, ERR-120, 85 p.

Clark, O.R., 1940, Interception of rainfall by prairie grasses, weeds, and certain crop plants: Ecological Monographs, v. 10, no. 2, p. 243.

DeFries, R.S., Foley, J.A., and Asner, G.P., 2004, Land-use choices—Balancing human needs and ecosystem function: Frontiers in Ecology and the Environment, v. 2, no. 5, p. 249–257.

Engstrom, D.R., Almendinger, J.E., and Wolin, J.A., 2009, Historical changes in sediment and phosphorus loading to the upper Mississippi River—Mass-balance reconstructions from the sediments of Lake Pepin: Journal of Paleolimnology, v. 41, no. 4, p. 563–588.

Fawcett, R.S., Christensen, B.R., and Tierney, D.P., 1994, The impact of conservation tillage on pesticide runoff into surface water: a review and analysis: Journal of Soil and Water Conservation, v. 49, no. 2, p. 126.

Flanagan, D.C., J.C.A. II, Nicks, A.D., Nearing, M.A., and Laflen, J.M., 1995, USDA—Water and Erosion Prediction Project *in* Flanagan, D.C., and Nearing, M.A., eds., Hillslope profile and watershed model documentation: West Lafayette, Ind., U.S. Department of Agriculture-Agricultural Research Service National Soil Erosion Research Laboratory, NSERL Report No. 10.

Foley, J.A., Kucharik, C.J., Twine, T.E., Coe, M.T., and Donner, S.D., 2004, Land use, land cover, and climate change across the Mississippi Basin—Impacts on selected land and water resources: Geophysical Monograph , v. 153, p. 249–261.

Food and Agriculture Organization of the United Nations (FAO), 2009, Major food and agricultural commodities and producers—Country rank in the world, by commodity: Food and Agriculture Organization of the United Nations, v. 2009, no. 04/03.

Foster, G.R., Lane, L.J., Nowlin, J.D., Young, J.M., and Laflen, R.A., 1981, Estimating erosion and sediment yield on field-sized areas: Transactions of the American Society of Agricultural Engineers, v. 24, no. 5, p. 1,253–1,262.

Ghidey, F., and Alberts, E., 1998, Runoff and soil losses as affected by corn and soybean tillage systems: Journal of Soil and Water Conservation, v. 53, no. 1, p. 64.

Gleason, R.A., and Euliss, N.H., 1983, Sedimentation of prairie pothole wetlands—The need for integrated research by agricultural and wildlife interests: Proceedings of the U.S. Committee of Irrigation and Drainage, v. 107, p. 114 p.

Gupta, S.C., and Larson, W.E., 1979, Estimating soil water retention characteristics from particle size distribution, organic matter percent, and bulk density: Water Resources Research, v. 15, no. 6, p. 1,633–1,635.

Henley, W.F., Patterson, M.A., Neves, R.J., and Lemly, A.D., 2000, Effects of sedimentation and turbidity on lotic food webs—A concise review for natural resource managers: Reviews in Fisheries Science, v. 8, no. 2, p. 125–139.

Huffman, R.L., Fangmeier, D.D., Elliot, W.J., Workman, S.R., and Schwab, G.O., 1993, Soil and water conservation engineering, 6th ed.: St. Joseph, MI: American Society of Agricultural and Biological Engineers.

Julien, P.Y., 1995, Erosion and sedimentation: New York, Cambridge University Press, 300 p.

Kent, K., 1973, A method for estimating volume and rate of runoff in small watersheds: U.S. Department of Agriculture, Soil Conservation Service Report SCS–TP–149, revised April 1973, 64 p.

Lin, H., Brooks, E., Mcdaniel, P., and Boll, J., 2008, Hydropedology and surface/subsurface runoff processes, *in* Anderson, M.G., and McDonnell, J.J., eds., Encyclopedia of hydrologic sciences: Hoboken, N.J., John Wiley and Sons, p. 25.

Mao, D., and Cherkauer, K.A., 2009, Impacts of land-use change on hydrologic responses in the Great Lakes region: Journal of Hydrology, v. 374, no. 1–2, p. 71–82.

Meyer, Chuck, 2004, General description of the CLIGEN model and its history: West Lafayette, Ind., U.S. Department of Agriculture-Agricultural Research Service, 21 p.

Montgomery, D.R., 2007, Soil erosion and agricultural sustainability: Proceedings of the National Academy of Sciences of the United States of America, v. 104, no. 33, p. 13,268–13,272.

Munn, M.D., Gilliom, R.J., Moran, P.W., and Nowell, L.H., 2006, Pesticide toxicity index for freshwater aquatic organisms (2d ed.): U.S. Geological Survey Scientific Investigations Report 2006–5148, 81 p.

Myers, J., 1996, Runoff and sediment loss from three tillage systems under simulated rainfall: Soil and Tillage Research, v. 39, no. 1–2, p. 115–129.

National Climatic Data Center, 2009, Daily climate records from 1948 through 2008 for COOP identification numbers 123527 and 458207: U.S. National Oceanic and Atmospheric Administration, accessed June 9, 2009, at http://www.ncdc noaa.gov/cdo-web/#t=secondTabLink.

Natural Resources Conservation Service, 2011, Soil texture calculator: U.S. Department of Agriculture Web site, accessed October 10, 2011, at http://soils.usda.gov/technical/aids/investigations/texture/.

Nemes, A., and Rawls, W.J., 2004, Soil texture and particle-size distribution as input to estimate soil hydraulic properties: Developments in Soil Science, v. 30, no. 4, p. 47–70.

Newcombe, C.P., and MacDonald, D.D., 1991, Effects of suspended sediments on aquatic ecosystems: North American Journal of Fisheries Management, v. 11, no. 1, p. 72–82.

Owoputi, L.O., and Stolte, W.J., 1995, Soil detachment in the physically based soil erosion process—A review: Transactions of the American Society of Agronomy, v. 38, no. 4, p. 1,099–1,110.

Pimentel, D., Allen, J., Beers, A., Guinand, L., Linder, R., McLaughlin, P., Meer, B., Musonda, D., Perdue, D., Poisson, S., and others, 1987, World agriculture and soil erosion: BioScience, v. 37, no. 4, p. 277–283.

Poff, N., Bledsoe, B., and Cuhaciyan, C., 2006, Hydrologic variation with land use across the contiguous United States—Geomorphic and ecological consequences for stream ecosystems: Geomorphology, v. 79, no. 3–4, p. 264–285.

Quinn, J.M., Davies-Colley, R.J., Hickey, C.W., Vickers, M.L., and Ryan, P.A., 1992, Effects of clay discharges on streams: Hydrobiologia, v. 248, no. 3, p. 235–247.

Scanlon, B.R., Reedy, R.C., Stonestrom, D.A., Prudic, D.E., and Dennehy, K.F., 2005, Impact of land use and land cover change on groundwater recharge and quality in the southwestern U.S.: Global Change Biology, v. 11, no. 10, p. 1,577–1,593.

Scharffenberg, W.A., and Fleming, M.J., 2010, Hydrologic Modeling System HEC-HMS—User's manual: Davis, Calif., U.S. Army Corps of Engineers.

Sisk, T., ed., 1999, Perspectives on the land-use history of North America—A context for understanding our changing environment: U.S. Fish and Wildlife Service Biological Science Report 1998–0003, 104 p.

Stone, J., Lane, L., and Shirley, E., 1992, Infiltration and runoff simulation on a plane: Transactions of the American Society of Agricultural Engineers, v. 35, no. 1, p. 161–170.

Strudley, M., Green, T., and Ascoughii, J., 2008, Tillage effects on soil hydraulic properties in space and time—State of the science: Soil and Tillage Research, v. 99, no. 1, p. 4–48.

Twine, T.E., Kucharik, C.J., and Foley, J.A., 2004, Effects of land cover change on the energy and water balance of the Mississippi River Basin: Journal of Hydrometeorology, v. 5, no. 4, p. 640–655.

U.S. Environmental Protection Agency, 2009, National Water Quality Inventory—Report to Congress, 2004 Reporting Cycle: Washington, D.C., U.S. Environmental Protection Agency, Office of Water, EPA 841–R–08–001, accessed June 22, 2012, at http://water.epa.gov/lawsregs/guidance/cwa/305b/2004report_index.cfm.

Unger, P.W., 1992, Infiltration of simulated rainfall—Tillage system and crop residue effects: Soil Science Society of America Journal, v. 56, no. 1, p. 283–289.

Uusitalo, R., Turtola, E., Kauppila, T., and Lilja, T., 2001, Particulate phosphorus and sediment in surface runoff and drainflow from clayey soils: Journal of Environmental Quality, v. 30, no. 2, p. 589–595.

van Dijk, A.I.J.M., and Bruijnzeel, L.A., 2001, Modelling rainfall interception by vegetation of variable density using an adapted analytical model—Part 1, model description: Journal of Hydrology, v. 247, no. 3–4, p. 230–238.

Ward, A.D., and Trimble, S.W., 2004, Environmental hydrology (2d ed.): Boca Raton, Fla., Lewis Publishers, 475 p.

Woolhiser, D.A., Smith, R.E., and Goodrich, D.C., 1990, KINEROS—A kinematic runoff and erosion model: U.S. Department of Agriculture-Agricultural Research Service, ARS–77, 130 p.

Yan, B., Tomer, M.D., and James, D.E., 2010, Historical channel movement and sediment accretion along the South Fork of the Iowa River: Journal of Soil and Water Conservation, v. 65, no. 1, p. 1–8.

Zhang, Y., and Schilling, K., 2006, Increasing streamflow and baseflow in Mississippi River since the 1940s—Effect of land use change: Journal of Hydrology, v. 324, no. 1–4, p. 412–422.

Appendix 1. Summary of Water Erosion Prediction Project Model Simulations

Table 1A.   Water Erosion Prediction Project model simulations of natural (pre-agricultural) land covers and the values of landscape variables on the unit field under a humid environment representative of Greenfield, Indiana, and an arid climate representative of Sunnyside, Washington, with sprinkler irrigation.

Variable	Simulation identifier	Simulation name	Land slope (percent)	Soil texture	Land cover/ crop type	Tillage type	Management/ modification
Land cover	P001	Prairie grass (humid)	3	Loam	Prairie grass	None	None
	P002	Forest (humid)	3	Loam	Forest	None	None
	P003	Scrub (arid)	3	Loam	Sagebrush	None	None

Table 1B.   Water Erosion Prediction Project model simulations and the values of landscape and agricultural variables on the unit field under a humid climate representative of Greenfield, Indiana.

Variable	Simulation identifier	Simulation name	Land slope (percent)	Soil texture	Land cover/ crop type	Tillage type	Management/ modification
Slope	H001	0.5 percent slope	0.5	Loam	Corn	Reduced–Spring	None
	H002	2.0 percent slope	2	Loam	Corn	Reduced–Spring	None
	H003[1]	3.0 percent slope	3	Loam	Corn	Reduced–Spring	None
	H004	6.0 percent slope	6	Loam	Corn	Reduced–Spring	None
	H005	8.0 percent slope	8	Loam	Corn	Reduced–Spring	None
	H006	10.0 percent slope	10	Loam	Corn	Reduced–Spring	None
Soil texture	H007	Sand	3	Sand	Corn	Reduced–Spring	None
	H008[1]	Loam	3	Loam	Corn	Reduced–Spring	None
	H009	Clay	3	Clay	Corn	Reduced–Spring	None
	H010	Silt	3	Silt	Corn	Reduced–Spring	None
	H011	Sandy clay	3	Sandy clay	Corn	Reduced–Spring	None
	H012	Sandy clay loam	3	Sandy clay loam	Corn	Reduced–Spring	None
	H013	Sandy loam	3	Sandy loam	Corn	Reduced–Spring	None
	H014	Loamy sand	3	Loamy sand	Corn	Reduced–Spring	None
	H015	Silt loam	3	Silt loam	Corn	Reduced–Spring	None
	H016	Silty clay loam	3	Silty clay loam	Corn	Reduced–Spring	None
	H017	Silty clay	3	Silty clay	Corn	Reduced–Spring	None
	H018	Clay loam	3	Clay loam	Corn	Reduced–Spring	None
Land cover/ crop type	P001	Humid prairie grass	3	Loam	Prairie grass	None	None
	P002	Humid forest	3	Loam	Forest	None	None
	H019	Fallow	3	Loam	None	Reduced–Spring	None
	H020	Fallow (untilled)	3	Loam	Fallow	None	None
	H021[1]	Corn	3	Loam	Corn	Reduced–Spring	None
	H022	Soybeans	3	Loam	Soybeans	Reduced–Spring	None
	H023	Wheat	3	Loam	Wheat	Reduced–Fall	None
	H024	Alfalfa	3	Loam	Alfalfa	Conventional–Spring	None
Tillage type	H025	No-till	3	Loam	Corn	No-Till	None
	H026[1]	Reduced–Spring	3	Loam	Corn	Reduced–Spring	None
	H027	Conservation–Fall	3	Loam	Corn	Conservation–Fall	None
	H028	Conventional–Fall	3	Loam	Corn	Conservation–Fall	None
	H029	Reduced–Fall	3	Loam	Corn	Reduced–Fall	None
	H030	Reduced-contour– Fall	3	Loam	Corn	Reduced-contour– Fall	None
Management/ modification	H031	None	3	Loam	Corn	Reduced–Spring	None
	H032	Strip cropping	3	Loam	Corn/wheat	Reduced–Spring	Strip cropping
	H033	Cover crop	3	Loam	Corn/rye	Reduced–Spring	Cover crop
	H034	Crop rotation	3	Loam	Corn/soybeans	Reduced–Spring	Crop rotation
	H035	Terraces	3	Loam	Corn	Reduced–Spring	Terraces
	H036	Subsurface drains	3	Loam	Corn	Reduced–Spring	Subsurface drains

[1]Base scenario reflects a common scenario of 3 percent slope with reduced-spring-till corn on a loam soil for aiding in uniform comparison across variable subsets.

Table 1C.   Water Erosion Prediction Project model simulations and the values of landscape and agricultural variables on the unit field under an arid climate with sprinkler irrigation representative of Sunnyside, Washington.

Variable	Simulation identifier	Simulation name	Land slope (percent)	Soil texture	Land cover/ crop type	Tillage type	Management/ modification
Slope	S001	0.5 percent slope	0.5	Loam	Corn	Reduced–Spring	None
	S002	2.0 percent slope	2	Loam	Corn	Reduced–Spring	None
	S003[1]	3.0 percent slope	3	Loam	Corn	Reduced–Spring	None
	S004	6.0 percent slope	6	Loam	Corn	Reduced–Spring	None
	S005	8.0 percent slope	8	Loam	Corn	Reduced–Spring	None
	S006	10.0 percent slope	10	Loam	Corn	Reduced–Spring	None
Soil texture	S007	Sand	3	Sand	Corn	Reduced–Spring	None
	S008[1]	Loam	3	Loam	Corn	Reduced–Spring	None
	S009	Clay	3	Clay	Corn	Reduced–Spring	None
	S010	Silt	3	Silt	Corn	Reduced–Spring	None
	S011	Sandy clay	3	Sandy clay	Corn	Reduced–Spring	None
	S012	Sandy clay loam	3	Sandy clay loam	Corn	Reduced–Spring	None
	S013	Sandy loam	3	Sandy loam	Corn	Reduced–Spring	None
	S014	Loamy sand	3	Loamy sand	Corn	Reduced–Spring	None
	S015	Silt loam	3	Silt loam	Corn	Reduced–Spring	None
	S016	Silty clay loam	3	Silty clay loam	Corn	Reduced–Spring	None
	S017	Silty clay	3	Silty clay	Corn	Reduced–Spring	None
	S018	Clay loam	3	Clay loam	Corn	Reduced–Spring	None
Land cover/ crop type	S019	Prairie grass	3	Loam	Prairie grass	None	None
	S020	Sagebrush	3	Loam	Sagebrush	None	None
	S021	Fallow	3	Loam	None	Reduced–Spring	None
	S022	Fallow (untilled)	3	Loam	Fallow	None	None
	S023[1]	Corn	3	Loam	Corn	Reduced–Spring	None
	S024	Soybeans	3	Loam	Soybeans	Reduced–Spring	None
	S025	Wheat	3	Loam	Wheat	Reduced–Fall	None
	S026	Alfalfa	3	Loam	Alfalfa	Conventional–Spring	None
Tillage type	S027	No-till	3	Loam	Corn	No-till	None
	S028[1]	Reduced–Spring	3	Loam	Corn	Reduced–Spring	None
	S029	Conservation–Fall	3	Loam	Corn	Conservation–Fall	None
	S030	Conventional–Fall	3	Loam	Corn	Conventional–Fall	None
	S031	Reduced–Fall	3	Loam	Corn	Reduced–Fall	None
	S032	Reduced-contour–Fall	3	Loam	Corn	Reduced contour–Fall	None
Management / modification	S033[1]	None	3	Loam	Corn	Reduced–Spring	None
	S034	Strip cropping	3	Loam	Corn/wheat	Reduced–Spring	Strip cropping
	S035	Cover crop	3	Loam	Corn/rye	Reduced–Spring	Cover crop
	S036	Crop rotation	3	Loam	Corn/soybeans	Reduced–Spring	Crop rotation
	S037	Terraces	3	Loam	Corn	Reduced–Spring	Terraces
	S038	Subsurface drains	3	Loam	Corn	Reduced–Spring	Subsurface drains

[1] Base scenario reflects a common scenario of 3 percent slope with reduced-spring-till corn on a loam soil for aiding in uniform comparison across variable subsets.

Appendix 2. Water Erosion Prediction Project Model Installation Files and Instructions

To reconstruct and perform any of the various simulations contained within this report, the WEPP model must be installed on a personal computer. The version of the WEPP model with a user interface for the Microsoft® Windows operating system and configuration files used for this report are in a compressed archive and are available at http://pubs.usgs.gov/sir/2012/5203/.

Once downloaded, create a directory named "WEPP" at a location on the local computer. Then decompress the ".zip" file, downloaded from the above URL to the WEPP directory on the local computer. To start the program, execute the "WinWEPP.exe" file in the "weppwin" subdirectory of the "WEPP" directory previously created.

To run any of the individual scenarios, select the "File" menu and select "Open Project Set." This will bring up a window containing the contents of a directory. Choose the subdirectory "WEPP_SIR_projects" and then select either "humid.prs" or "arid.prs." Finally, select the "project ID's" corresponding to the simulations of interest (appendix 1).Then click "Run Selected Projects" from the WinWEPP interface.

To determine the water budget components from a simulation, a WEPP output file named "XXXX_grph.txt" must be created and imported into a spreadsheet or similar computationally capable software: where XXXX refers to the simulation identifier (ID) (see appendix 1). To create these files, the "Keep Output Files" tick box must be checked near the top of the "WinWEPP project" screen. These files contain more than 100 parameters on a daily basis for the entire WEPP simulation and can therefore be large file sizes. To compute water budget and sediment yields for specific time frames the columns of interest in these files are: Day number (Column 1), Precipitation in millimeters per day (Column 2), Irrigation Depth in millimeters per day (Column 14), Runoff in millimeters (Column 16), Evapotranspiration in millimeters per day (Column 43), Drainage flux in meters per day (Column 44), Seepage in millimeters per day (Column 56), Total soil water in millimeters (Column 59), Snow depth in millimeters (Column 74), snow density in kilograms per square meter (Column 76) , and Total frozen soil water in millimeters (Column 94).

To perform a water balance from day i through day j, use the following formula over the range of rows on the appropriate water budget component columns:

Equation A.I: Water balance over an indefinite period of time

$$\sum_i^j P + I = \sum_i^j (RO + R + ET + 1000 * Dr) + \Delta S_{i,j} \qquad \text{(A.I)}$$

where

P is daily precipitation depth,

I is daily irrigation depth,

RO is daily runoff depth,

R is daily recharge / seepage depth,

ET is the daily depth of evapotranspiration,

Dr depth water drained by subsurface drains :
note conversion of meters per day to
millimeters per day, and

$\Delta S_{i,j}$ is the net change in water storage over the period
i through j (see following equation).

Equation A.II: Expansion of change in water storage term from Equation A.I

$$\Delta S_{i,j} = \left(SW_j - SW_i \right) + \frac{\left(SnD_j * \rho Sn_j - SnD_i * \rho Sn_i \right)}{pW} + \left(FrSW_j - FrSW_i \right) \qquad \text{(A.II)}$$

where,

$\Delta S_{i,j}$ is the net change in water storage over the
period i through j,

SW is the depth of soil water on days i and j,

SnD is the depth of snow on days i and j,

ρSn is the density of snow on days i and j,

pW is the density of water: 100 kilograms per cubic
meter (Chow and others, 1988), and

$FrSW$ is the depth of frozen soil water on days i and j.

Appendix 3. Climate and Irrigation Data Used in Simulations

The climate and irrigation data used for the simulations contained in this report are available for download at http://pubs.usgs.gov/sir/2012/5203/.

Appendix 4. Annualized Summaries of Water Budget Components for Simulations

Annual summary data of water budget components and sediment yields for each simulation are in comma separated value format files within a compressed archive and are available for download at http://pubs.usgs.gov/sir/2012/5203/. The files contain cumulative annual values and statistical summaries for each 60-year simulation. See appendix 1 for reference on simulation identifiers.

www.ingramcontent.com/pod-product-compliance
Lightning Source LLC
Chambersburg PA
CBHW081624170526
45166CB00009B/3092